ASSESSMENT OF TECHNOLOGIES DEPLOYED TO IMPROVE AVIATION SECURITY

First Report

Panel on Assessment of Technologies Deployed to Improve Aviation Security

National Materials Advisory Board
Commission on Engineering and Technical Systems
National Research Council

Publication NMAB-482-5
National Academy Press
Washington, D.C.

NATIONAL ACADEMY PRESS **2101 Constitution Avenue, N.W.** **Washington, D.C. 20418**

NOTICE: The project that is the subject of this report was approved by the Governing Board of the National Research Council, whose members are drawn from the councils of the National Academy of Sciences, the National Academy of Engineering, and the Institute of Medicine. The members of the committee responsible for the report were chosen for their special competencies and with regard for appropriate balance.

This report has been reviewed by a group other than the authors according to procedures approved by a Report Review Committee consisting of members of the National Academy of Sciences, the National Academy of Engineering, and the Institute of Medicine.

The National Academy of Sciences is a private, nonprofit, self-perpetuating society of distinguished scholars engaged in scientific and engineering research, dedicated to the furtherance of science and technology and to their use for the general welfare. Upon the authority of the charter granted to it by the Congress in 1863, the Academy has a mandate that requires it to advise the federal government on scientific and technical matters. Dr. Bruce Alberts is president of the National Academy of Sciences.

The National Academy of Engineering was established in 1964, under the charter of the National Academy of Sciences, as a parallel organization of outstanding engineers. It is autonomous in its administration and in the selection of its members, sharing with the National Academy of Sciences the responsibility for advising the federal government. The National Academy of Engineering also sponsors engineering programs aimed at meeting national needs, encourages education and research, and recognizes the superior achievements of engineers. Dr. William Wulf is president of the National Academy of Engineering.

The Institute of Medicine was established in 1970 by the National Academy of Sciences to secure the services of eminent members of appropriate professions in the examination of policy matters pertaining to the health of the public. The Institute acts under the responsibility given to the National Academy of Sciences by its congressional charter to be an adviser to the federal government and, upon its own initiative, to identify issues of medical care, research, and education. Dr. Kenneth I. Shine is President of the Institute of Medicine.

The National Research Council was organized by the National Academy of Sciences in 1916 to associate the broad community of science and technology with the Academy's purposes of furthering knowledge and advising the federal government. Functioning in accordance with general policies determined by the Academy, the Council has become the principal operating agency of both the National Academy of Sciences and the National Academy of Engineering in providing services to the government, the public, and the scientific and engineering communities. The Council is administered jointly by both Academies and the Institute of Medicine. Dr. Bruce Alberts and Dr. William Wulf are chairman and vice chairman, respectively, of the National Research Council.

This study by the National Materials Advisory Board was conducted under Contract No. DTFA03-94-C-00068 with the Federal Aviation Administration. Any opinions, findings, conclusions, or recommendations expressed in this publication are those of the author(s) and do not necessarily reflect the views of the organizations or agencies that provided support for the project.

Available in limited supply from:
National Materials Advisory Board
2101 Constitution Avenue, N.W.
HA-262
Washington, D.C. 20418
202-334-3505
nmab@nas.edu

Additional copies are available for sale from:
National Academy Press
Box 285
2101 Constitution Ave., N.W.
Washington, D.C. 20055
800-624-6242
202-334-3313 (in the Washington metropolitan area)
http://www.nap.edu

International Standard book Number: 0-309-06787-1

Printed in the United States of America.

PANEL ON ASSESSMENT OF TECHNOLOGIES DEPLOYED TO IMPROVE AVIATION SECURITY

THOMAS S. HARTWICK (chair), consultant, Seattle, Washington
ROBERT BERKEBILE, consultant, Leesburg, Florida
HOMER BOYNTON, consultant, Hilton Head Island, South Carolina
BARRY D. CRANE, Institute for Defense Analyses, Alexandria, Virginia
COLIN DRURY, State University of New York at Buffalo
LEN LIMMER, consultant, Fort Worth, Texas
HARRY E. MARTZ, Lawrence Livermore National Laboratory, Livermore, California
JOSEPH A. NAVARRO, JAN Associates, Bethesda, Maryland
ERIC R. SCHWARTZ, The Boeing Company, Seattle, Washington
ELIZABETH H. SLATE, Cornell University, Ithaca, New York
MICHAEL STORY, Thermo Instruments Systems, Santa Clara, California

Technical Consultants

RODGER DICKEY, Dallas-Fort Worth Airport Authority, Dallas, Texas
MOHSEN SANAI, SRI International, Menlo Park, California

National Materials Advisory Board Liaison

JAMES WAGNER, Case Western Reserve University, Cleveland, Ohio

National Materials Advisory Board Staff

SANDRA HYLAND, senior program manager (until June 1998)
CHARLES T. HACH, staff officer
JANICE M. PRISCO, project assistant
RICHARD CHAIT, NMAB director

NATIONAL MATERIALS ADVISORY BOARD

Preface

This is the first of four reports assessing the deployment of technologies (i.e., equipment and procedures) by the Federal Aviation Administration (FAA). This assessment of the 1997–1998 deployment of technologies by the FAA to improve aviation security was conducted by the Panel on Assessment of Technologies Deployed to Improve Aviation Security under the auspices of the National Research Council (NRC) Committee on Commercial Aviation Security. This is the first part of a four-part assessment that will be completed in fiscal year 2001. The subsequent parts of this study will be continued by a new committee that will be convened by the NRC in 1999. The form of this report reflects the panel's understanding of this study as part of a larger project and carefully distinguishes the issues and topical areas that could be completed in the first year from those that would require further study.

Based on the experience of the Committee on Commercial Aircraft Security and in anticipation of further queries from the FAA or other government entities deliberating on the continuation and deployment of equipment purchases in the coming fiscal year, the panel has endeavored to make a rapid assessment and generate a timely report in 1999. Therefore, the panel considered the major issues and overall effectiveness of the deployed technologies, postponing detailed descriptions and detailed discussions of less urgent topics until later. The panel was greatly assisted by the cooperation of the FAA, the U.S. Department of Transportation (DOT), and several airport and airline officials.

APPROACH AND SCOPE OF THIS STUDY

This study was conducted in response to a congressional directive (Section 303 PL 104-264, 1996) that the FAA engage the NRC to study the deployment of airport security equipment. The FAA requested that the NRC—the operating arm of the National Academy of Sciences—assess the operational performance of explosives-detection equipment and hardened unit-loading devices (HULDs) in airports and compare it to performance in laboratory testing to determine how to deploy this equipment more effectively to improve aviation security. As requested by Congress, the study was intended to address the following issues:

1. Assess the weapons and explosives-detection technologies available at the time of the study that are capable of being effectively deployed in commercial aviation.
2. Determine how the technologies referred to in paragraph (1) could be used more effectively to promote and improve security at airport and aviation facilities and other secured areas.
3. Assess the cost and advisability of requiring hardened cargo containers to enhance aviation security and reduce the required sensitivity of bomb-detection equipment.
4. On the basis of the assessments and determinations made under paragraphs (1), (2), and (3), identify the most promising technologies for improving the efficiency and cost effectiveness of weapons and explosives detection.

The NRC responded by convening the Panel on Assessment of Technologies Deployed to Improve Aviation Security, under the auspices of the Committee on Commercial Aviation Security of the National Materials Advisory Board. Interpretation of the four points presented by Congress and subsequent discussions between the FAA and the NRC led to the panel being asked to complete the following tasks:

1. Review the performance in laboratory tests of the explosives-detection technologies selected for deployment by the FAA's Security Equipment Integrated Product Team (SEIPT).
2. Assess the performance of the explosives-detection equipment deployed in airports in terms of detection capabilities, false-alarm rates, alarm resolution, operator effectiveness, and other operational aspects.

3. Recommend further research and development that might lead to reduced false-alarm rates and improved methods of alarm resolution.
4. Recommend methods of improving the operational effectiveness of explosives-detection equipment already deployed or about to be deployed in airports.
5. Assess different combinations of explosives-detection equipment and recommend ways to improve their effectiveness.
6. Review and comment on the FAA's plans for gathering metrics on field performance based on certification requirements of the explosives-detection equipment.
7. Assess the effectiveness of combining passenger profiling and passenger-bag matching with explosives-detection techniques.
8. Review the technical approach used to develop hardened aviation-cargo containers.
9. Review the results of tests of hardened cargo containers that have been used operationally by the air carriers.
10. Assess the overall operational experiences of air carriers in deploying hardened cargo containers.
11. Recommend scenarios for implementing hardened cargo containers to complement other aviation security measures, such as the deployment of explosives-detection equipment and passenger profiling.
12. Recommend further research and development that might lead to more effective hardened cargo containers.

Since this is the first of four reports assessing the FAA's deployment of technologies to improve aviation security, not every task item is fully addressed in this report. Furthermore, it is difficult to state definitively to what degree each individual task has been covered in this report because information obtained during the continuation of this study may lead to the task being revisited and/or revised in a later report. In this report, the panel has addressed, at least in part, tasks 1, 2, 3, 4, 6, 7, 11, and 12.

METHODOLOGY

The Panel on Assessment of Technologies Deployed to Improve Aviation Security developed this report based on: (1) panel meetings and technical literature provided by the FAA and the NRC staff; (2) presentations by outside experts on explosives-detection technologies, HULDs, passenger profiling, bag matching, airport-flow models, and the status of the deployment of equipment and implementation of security procedures; and (3) site visits by select panel members to John F. Kennedy International Airport, Los Angeles International Airport, San Francisco International Airport, and the FAA HULD test facility in Tucson, Arizona. Several factors were used in selecting these airports for site visits. All three are large "Category X" airports with international flights. Because Category X airports were the first to receive explosives-detection equipment, the panel was assured that the equipment would be operating and available for viewing. Because of the size of these airports, the panel was able to see deployed equipment in different installation configurations at one airport. During these visits, the panel studied the configurations of the deployed equipment and interviewed equipment operators and other security and baggage-handling employees.

Some panel members were invited to visit the FAA Technical Center in Atlantic City, New Jersey, and InVision Technologies in San Francisco, California, and to attend the Society of Automotive Engineers (SAE) meeting on air cargo and ground equipment in New Orleans, Louisiana. Finally, some members of the panel participated in a conference call with representatives of domestic air carriers. All panel members were selected for their expertise in technologies for explosives detection, operational testing, human factors and testing, structural materials and design, and air carrier and airport operations and design.

Panel Meetings

The panel met four times between January and August 1998 to gather information for this report. In the course of these meetings, the panel received briefings and reviewed technical literature on various aspects of security technologies and their deployment. Information was provided by experts from the FAA, as well as by outside experts.

Site Visits

A group of panel members visited San Francisco International Airport, John F. Kennedy International Airport, and Los Angeles International Airport to observe the operation of security equipment, including the FAA-certified InVision CTX-5000, several trace explosives-detection devices, and noncertified bulk explosives-detection equipment. Panel members were also able to meet with personnel from the airlines, airports, and private security contractors to discuss baggage handling, the use of containers, and security procedures. Local FAA personnel were also available to answer questions. Following the site visit to San Francisco International Airport, the panel members visited InVision Technologies in Newark, California, where they were informed of InVision's technical objectives and planned improvements to their explosives-detection systems.

In addition to the airport site visits, one panel member attended the FAA test of the Galaxy HULD, which passed the FAA blast criterion. This test took place at the FAA test facility in Tucson, Arizona. This visit provided a firsthand account of the FAA's test procedures and test results and provided an opportunity for a panel member to interact with members of the HULD design team. This site visit was followed by attendance at an SAE meeting on air cargo and ground equipment, at which current and former airline representatives, designers, and engineers described their perspectives on the potential deployment of HULDs. Finally, one panel member visited the FAA Technical Center to discuss human-factors issues pertaining to the deployment and operation of bulk and trace explosives-detection equipment. During this visit, the panel member was informed of progress on the development of the threat image projection system for testing operators of security equipment.

PHILOSOPHY

The deployment of security equipment is not just a technical issue or an airport operations issue or a funding issue. Effective deployment is a complex systems-architecture issue that involves separate but intertwined technical, management, funding, threat, and deployment issues. The panel was unanimous in its characterization of deployment as a total systems architecture and in its agreement to conduct this study from that perspective. This systems approach is the foundation of this report.

Thomas S. Hartwick, chair
Panel on Assessment of Technologies
Deployed to Improve Aviation Security

Acknowledgments

The Panel on Assessment of Technologies Deployed to Improve Aviation Security would like to acknowledge the individuals who contributed to this study, including the following speakers: Michael Abkin, ATAC; Jean Barrette, Transport Canada; Leo Boivan, Federal Aviation Administration; Jay Dombrowski, Northwest Airlines; Tony Fainberg, Federal Aviation Administration; Cathal Flynn, Federal Aviation Administration; Frank Fox, Federal Aviation Administration; Dwight Fuqua, TRW; Ken Hacker, Federal Aviation Administration; Trish Hammar, DSCI; Mike McCormick, Federal Aviation Administration; James Padgett, Federal Aviation Administration; Ron Pollilo, Federal Aviation Administration; Fred Roder, Federal Aviation Administration; Roshni Sherbondi, Federal Aviation Administration; and Alexis Stefani, U.S. Department of Transportation.

The panel is also grateful for the contributions of the contracting office technical representatives, Paul Jankowski and Alan K. Novakoff. In addition, the panel is appreciative of the insights provided by Nelson Carey, Federal Aviation Administration; John Daly, U.S. Department of Transportation; Howard Fleisher, Federal Aviation Administration; Lyle Malotky, Federal Aviation Administration; Ronald Polillo, Federal Aviation Administration; and Ed Rao, Federal Aviation Administration.

This report has been reviewed in draft form by individuals chosen for their diverse perspectives and technical expertise, in accordance with procedures approved by the NRC's Report Review Committee. The purpose of this independent review is to provide candid and critical comments that will assist the institution in making the published report as sound as possible and to ensure that the report meets institutional standards for objectivity, evidence, and responsiveness to the study charge. The review comments and draft manuscript remain confidential to protect the integrity of the deliberative process. We wish to thank the following individuals for their participation in the review of this report: Jon Amy, Purdue University; Albert A. Dorman, AECOM; Michael Ellenbogen, Vivid Technologies; Arthur Fries, Institute for Defense Analyses; Valerie Gawron, Calspan; James K. Gran, SRI International; Robert E. Green, Johns Hopkins University; John L. McLucas, Consultant; Hyla Napadensky, Napadensky Energetics (retired); Robert E. Schafrik, GE Aircraft Engines; and Edward M. Weinstein, Galaxy Scientific Corporation. While the individuals listed above have provided constructive comments and suggestions, it must be emphasized that responsibility for the final content of this report rests entirely with the authoring committee and the NRC.

For organizing panel meetings and directing this report to completion, the panel would like to thank Charles Hach, Sandra Hyland, Lois Lobo, Janice Prisco, Shirley Ross, Teri Thorowgood, and Pat Williams, staff members of the National Materials Advisory Board. The panel is also appreciative of the efforts of Carol R. Arenberg, editor, Commission on Engineering and Technical Systems.

Contents

Tables, Figures, and Boxes

TABLES

FIGURES

BOXES

Acronyms and Abbreviations

ATA	Air Transport Association
CAPS	computer-assisted passenger screening
CT	computed tomography
DF	detection first
EDS	explosives-detection system
FAA	Federal Aviation Administration
FAR	Federal Aviation Regulation
HULD	hardened unit-loading device
IATA	International Air Transport Association
ICAO	International Civil Aviation Organization
MBTS	modular bomb test set
NRC	National Research Council
O&S	operational and support
PPBM	positive passenger-bag matching
PRA	probabilistic risk assessment
SAE	Society of Automotive Engineers
SEF	security enhancement factor
SEIPT	Security Equipment Integrated Product Team
SOS	system of systems
ST	simulated terrorist
TAAS	total architecture for aviation security
TEDD	trace explosives-detection device
TEDDCS	TEDD calibration standard
TIPS	threat image projection system
TPF	throughput first
ULD	unit-loading device
P_d	probability of detection
P_{fa}	probability of false alarm

Executive Summary

In 1997, the Federal Aviation Administration (FAA) was directed by President Clinton and authorized by Congress (PL 104-264, PL 104-208) to deploy 54 FAA-certified explosives-detection systems[1] (EDSs) and more than 400 trace explosives-detection devices (TEDDs) at airports around the country. The purpose of these deployments was to prevent attacks against civil aviation, such as the recent attempt described in Box ES-1. This report, which assesses the FAA's progress in deploying and utilizing equipment and procedures to enhance aviation security, was produced by the National Research Council (NRC) in response to a congressional directive to the FAA (PL 104-264 § 303). This is the first of four reports assessing the deployment of technologies (i.e., equipment and procedures) by the FAA. In this report the 1997–1998 deployment of technologies by the FAA to improve aviation security is assessed. This panel was convened under the auspices of the NRC Committee on Commercial Aviation Security. Although appropriations are authorized for this assessment through fiscal year 2001, the Committee on Commercial Aviation Security will conclude its work in 1999. Therefore, with the agreement of the FAA, the assessment will be continued by a new committee that will be convened by the NRC in 1999. The form of this report reflects the panel's understanding of this study as part of an ongoing assessment (based on the enabling congressional language). For this reason, the panel carefully distinguished issues and topical areas that could be completed in the first year from those that would require further study.

This report assesses the operational performance of explosives-detection equipment and hardened unit-loading devices (HULDs) in airports and compares their operational performance to their laboratory performance, with a focus on improving aviation security. As requested by Congress, this report addresses (in part) the following issues:

1. Assess the weapons and explosive-detection technologies available at the time of the study that are capable of being effectively deployed in commercial aviation.
2. Determine how the technologies referred to in paragraph (1) could be used more effectively to promote and improve security at airport and aviation facilities and other secured areas.
3. Assess the cost and advisability of requiring hardened cargo containers to enhance aviation security and reduce the required sensitivity of bomb-detection equipment.
4. On the basis of the assessments and determinations made under paragraphs (1), (2), and (3), identify the most promising technologies for improving the efficiency and cost effectiveness of weapons and explosives detection.

This panel considers aviation security as a total system architecture and measures the effectiveness of deployment on that basis.

DEPLOYED TECHNOLOGIES

The congressionally mandated deployment of bulk explosives-detection equipment began in January 1997 and continued throughout 1998. The FAA formed the Security Equipment Integrated Product Team (SEIPT) to carry out this deployment. The SEIPT assessed the availability of explosives-detection equipment capable of being effectively deployed in commercial aviation and formulated a plan to deploy this equipment in airports throughout the United States. In a separate program, the FAA has tested HULDs designed to contain a discrete explosive blast. Ten HULDs

[1] The following terminology is used throughout this report. An *explosives-detection system* is a self-contained unit composed of one or more integrated devices that has passed the FAA's certification test. An *explosives-detection device* is an instrument that incorporates a single detection method to detect one or more explosive material categories. *Explosives-detection equipment* is any equipment, certified or otherwise, that can be used to detect explosives.

BOX ES-1
A Recent Attempt to Attack U.S. Commercial Aircraft

On April 22, 1995, FBI agents took custody from Philippine authorities of Abdul Hakim Murad. Murad was arrested after a fire broke out in a Manila apartment in which he, Ramzi Yousef, and another associate were living and where officials found explosives and bomb-making materials (FBI, 1995). This fire may well have prevented the worst terrorist attack against civil aviation in history. Yousef was later indicted for the 1994 bombing of Philippine Airline Flight 434, which was determined to be a test run of a plot to blow up 11 American planes simultaneously (Zuckerman, 1996). Although it is horrifying to contemplate what might have happened if a fire had not broken out in Murad's apartment, it is more constructive to focus on what has been done—and what is being done—to improve aviation security.

have been deployed to three air carriers for operational testing.

The FAA's aviation security equipment and procedures include bulk[2] explosives-detection equipment, TEDDs, HULDs, computer-aided passenger screening (CAPS), and positive passenger-bag matching (PPBM). These equipment and procedures are described in Table ES-1.

FINDINGS

It is well documented (e.g., GAO 1998; DOT, 1998) that the FAA/SEIPT is behind schedule in the deployment of aviation security equipment. In 1997, Congress provided $144.2 million for the purchase of commercially available screening equipment, and the FAA/SEIPT planned to deploy 54 certified EDSs and 489 TEDDs by December 1997 (GAO, 1998). In addition, the FAA planned to implement CAPS fully by December 1997. Once it became apparent that these goals could not be met, the FAA set a new goal of deploying 54 certified EDSs, 22 noncertified bulk explosives-detection devices, and 489 TEDDs by December 31, 1998. The FAA also planned to implement CAPS fully by December 31, 1998. As of January 1, 1999, 71 certified EDSs, six noncertified bulk explosives-detection devices, and 366 TEDDs had been installed in airports, and CAPS and PPBM had been adopted by six airlines. In addition, 10 HULDs have been deployed to three airlines for operational testing.

The panel concluded that the combined efforts of the government, the airlines, and the airports to date have been effective in deploying aviation security technologies (improving aviation security to a level that will be quantified when additional data are collected during future studies), although, because of the urgent need for immediate action against incipient terrorism (White House Commission on Aviation Safety and Security, 1997), equipment and procedures were implemented rapidly without regard for how they would contribute to a total architecture for aviation security (TAAS). The panel believes that definition of such an architecture is essential to the success of this program; hence, it suggests formality in defining and using a TAAS. That is, although the capacity of individual pieces of equipment to discretely improve security at the point of deployment is known to some degree, the integrated effect of the total deployment of equipment and the implementation of procedures on the whole of aviation security is not. After much deliberation, the panel concluded that the performance of the TAAS could be measured by a single factor, the security enhancement factor (SEF), which will enable a quantitative evaluation of the performance of diverse deployment scenarios and show the importance of specific elements (e.g., explosives-detection equipment) to the performance of the TAAS.

RESPONSE TO CONGRESS

Protecting civil aviation against terrorist threats is a complex problem. Given the short response time and the complexity of the terrorist threat, the panel concluded that the research, development, and deployment by the FAA and others have been successful in qualitative terms. The urgent need for security equipment and procedures, expressed by the White House Commission on Aviation Security and Safety and by Congress in 1997, did not leave time for extensive system analyses. Therefore, the FAA proceeded with the deployment of hardware as it became available. Hence, the *security system* has evolved as the hardware has become available. It is not surprising, therefore, that data describing the efficacy of the deployed equipment are inadequate. The lack of performance data and the incomplete integration of the equipment into a complete security architecture are issues that any large system developer would be likely to encounter at this stage of development. The absence of a system architecture is the basis for the major recommendations of the panel. Nevertheless, the FAA will have to address these issues in the future.

Explosives-detection equipment and HULDs are part of a total system architecture and should be evaluated in the context of a TAAS. Although the FAA, its contractors, the airlines, and the airports have adopted some elements of the

[2]In this report, *bulk explosives* include all forms and configurations of an explosive at threat level (e.g., shaped explosives, sheet explosives, etc.).

TABLE ES-1 Selected Aviation Security Equipment and Procedures

Technology	Description
Computer-assisted passenger screening (CAPS)	CAPS is a system that utilizes a passenger's reservation record to determine whether the passenger can be removed from consideration as a potential threat. If the passenger cannot be cleared (i.e., determined not to be a threat), CAPS prompts the check-in agent to request additional information from the passenger for further review. If this information is still insufficient to clear the passenger, the passenger's bags and the passenger are considered "selectees" and are routed through additional security procedures.
Positive passenger-bag match (PPBM)	PPBM is a security procedure that matches the passenger's checked baggage with the passenger to ensure that baggage is not loaded aboard an airplane unless the passenger also boards. This security measure is implemented for all outbound international flights and for some domestic flights.
FAA-certified explosives detection systems (EDSs)	An EDS is a self-contained unit composed of one or more integrated explosives-detection devices that have passed the FAA's certification test. As of April 1999 only computed-tomography-based technologies have passed the FAA bulk explosives-detection certification tests (e.g., InVision CTX-5000, CTX-5000 SP, and CTX-5500 DS).
Bulk explosives-detection equipment	Bulk explosives-detection equipment includes any explosives-detection device or system that remotely senses some physical or chemical property of an object under investigation to determine if it is an explosive. This equipment, primarily used for checked baggage, consists of quadrupole resonance and advanced x-ray technologies, including radiography and tomography.
Trace explosives-detection devices (TEDDs)	TEDDs involve the collection of particles or vapor from the object under investigation to determine if an explosive is present. TEDDs are being deployed for several threat vectors: carry-on baggage (especially electronic devices), passengers, checked baggage, and cargo. TEDDs employ a variety of techniques for detecting vapors, particles, or both, which include chemiluminescence, ion mobility spectroscopy, and gas chromatography. TEDDs do not indicate the amount of explosive present and hence do not reveal the presence of a bomb, except inferentially.
Hardened unit-loading devices (HULDs)	A HULD is a specially designed baggage container that can contain the effects of an internal explosion without causing damage to the aircraft. A design by Galaxy Scientific passed the FAA blast test in March 1998. A second Galaxy Scientific design passed the FAA blast test in January 1999. To study operational performance and reliability, the FAA deployed 10 Galaxy HULDs in 1999.

total systems approach, in the panel's opinion they have not gone far enough. This study, and future aviation security studies conducted by the NRC, will be most useful to the FAA if they adopt the recommended comprehensive TAAS approach. Furthermore, adopting the TAAS approach will enable the FAA (and others) to characterize improvements in aviation security quantitatively using the SEF.

The panel has addressed (in part) the four points raised by Congress below. For clarity these points are listed again, followed by the relevant conclusions and recommendations.

1. **Assess the weapons and explosives-detection technologies available at the time of the study that are capable of being effectively deployed in commercial aviation.**

This study focused on explosives-detection technologies. While it is conceivable that some of these technologies could also be used for weapons detection, this topic was not addressed in this report.

Bulk Explosives-Detection Equipment

The vast majority of bulk explosives-detection equipment deployed is the FAA-certified InVision CTX-series EDS (explosives-detection system). Most of the performance data on this equipment was generated during laboratory testing—largely certification testing—at the FAA Technical Center. Certification tests, however, only reflect the ability of the equipment to detect a bag that contains an explosive, and the detection rates are based on bag-alarm rates. That is, an explosive is considered to be detected if the alarm is set off for the bag containing the explosive, even if the alarm is triggered by a nonexplosive object in the bag. Certification testing does not measure alarm resolution and does not include testing in the operational environment of an airport, making it difficult to assess explosives-detection technologies for deployment. In the panel's opinion, some of the unanticipated problems encountered with the CTX-5000 SP in the field can be reasonably related to the limitations of certification testing. Under current certification guidelines, equipment certified in the future may encounter similar problems.

Recommendation. During certification testing, the FAA should, whenever possible, measure both true detection rates (i.e., correctly identifying where an explosive is when an alarm occurs), and false-detection rates (i.e., an alarm triggered by something other than an explosive in a bag that contains an explosive). The FAA should also include the ability of explosives-detection equipment to assist operators in resolving alarms (including in an airport) as part of certification testing. Alarm resolution should be included in the measurement of throughput rate, detection rate, and false-alarm rate.

Trace Explosives-Detection Devices

TEDDs are widely used in airports, but no comprehensive methodology has been developed to evaluate their effectiveness, such as standard test articles or instrument and operator requirements. Because no standard test articles for TEDDs have been demonstrated—and because of the resultant inability to separate instrument and operator performance—it is not possible to measure the performance of TEDDs.

Recommendation. The FAA should develop and implement a program to evaluate the effectiveness of deployed trace explosives-detection devices. This evaluation should include measurements of instrument and operator performance, including measurements in the deployed (i.e., airport) environment.

Computer-Assisted Passenger Screening and Positive Passenger-Bag Matching

CAPS appears to be an effective way to screen passengers to identify selectees who require further security measures, such as bag matching or bag screening. The panel anticipates that PPBM combined with CAPS will be an effective tool for improving aviation security. Despite the positive attributes of CAPS, the panel is concerned that the FAA has not demonstrated a measure for characterizing quantitatively the effectiveness of CAPS. A CAPS selectee could bypass PPBM by checking a bag at the gate or the door of the aircraft (as opposed to the ticket counter). Furthermore, PPBM has not been demonstrated to be effective when a selectee changes planes at a connecting airport. That is, passengers identified as selectees at originating airports (who are then subject to PPBM) are not subject to PPBM on subsequent connections of that flight. Another shortfall of PPBM is when a passenger checks a bag (or bags) at the gate.

Recommendation. Computer-assisted passenger screening (CAPS) should continue to be used as a means of identifying selectee passengers whose bags will be subject to positive passenger-bag matching (PPBM), screening by explosives-detection equipment, or both. PPBM combined with CAPS should be part of the five-year plan recommended below. Passengers designated as selectees at the origination of their flights should remain selectees on all connecting legs of their flights. Within six months, the FAA should develop and implement a method of testing the effectiveness of CAPS.

2. **Determine how the technologies referred to in paragraph (1) could be used more effectively to promote and improve security at airport and aviation facilities and other secured areas.**

Progress in the Deployment of Aviation Security Equipment

The panel concluded that the FAA/SEIPT, the airlines, airports, and associated contractors have gained significant experience from the initial deployment of security equipment and procedures, and the current implementation of security equipment does not appear to have interfered unreasonably with airline operations. Most importantly, in the collective opinion of the panel, the deployment of security equipment has improved aviation security. The panel believes that continued emphasis on, funding of, and deployment of security equipment will further enhance aviation security. Future deployments should be more efficient if they are based on the experience from the initial deployment.

Recommendation. The U.S. Congress should continue to fund and mandate the deployment of commercially available explosives-detection equipment through the FAA/SEIPT. Continued deployments will increase the coverage of domestic airports and eventually provide state-of-the-art security equipment systemwide. Further deployments can improve aviation security in the short term and provide the infrastructure for mitigating potential threats in the long term.

Operator Performance

Human operators are integral to the performance of all deployed explosives-detection equipment. Because fully automated explosives-detection equipment will not be developed in the foreseeable future, particularly with respect to alarm resolution, human operators will continue to be immensely important to realizing the full potential of deployed security hardware. The TAAS analysis presented in this report quantifies the impact of the operator on the SEF. Certification testing of explosives-detection equipment, however, does not include testing of human operators. Current testing only defines the operational capability (or performance) of the equipment.

Recommendation. The FAA should institute a program to qualify security-equipment operators to ensure that the human operator/explosives-detection system (EDS) combination meets the performance requirements of a certified EDS. This program should include the definition of operator performance standards and a means of monitoring operator performance. The FAA should implement this program within six months of receipt of this report.

Measuring Operational Performance

Because of the paucity of operational data for deployed explosives-detection equipment, the panel found it impracticable to characterize the deployment status of security equipment and processes quantitatively. The data are insufficient both for the equipment and for operator performance, and no quantitative measures of the effectiveness of the total security system (e.g., TAAS) were provided to the panel. The majority of data focused on subsystems, such as bulk explosives-detection systems. A thorough assessment of equipment and system performance requires well defined performance metrics and the collection of data. The panel concluded that the FAA has not defined adequate performance metrics for security subsystems (e.g., TEDDs) or for the TAAS.

Recommendation. The FAA should make a concerted effort to define operational performance metrics for security subsystems and for the total architecture for aviation security (TAAS). The FAA should also create an action team in the next six months to systematically collect operational data, which should be used to optimize the TAAS, as well as to identify and correct substandard performance of equipment and operators. The data collected would also provide insights into the deployment and use of equipment in the future.

Measuring Security Enhancement

Besides the dearth of operational data and total-system performance metrics, the FAA has not defined an overall measure of security enhancement. The primary performance measure for the TAAS is, of course, protection against the threat of explosives. Consequently, the panel believes the critical factor in assessing the performance of the TAAS is the measure of false negatives (i.e., unidentified bags that contain explosives). The panel defined improved performance (i.e., the SEF) as the ratio of the number of simulated bombs that defeat the baseline security system to the number of simulated bombs that defeat the newly deployed system.

Recommendation. The FAA should formulate a security enhancement factor (SEF) for the integrated total architecture for aviation security systems. The SEF should be calculated from data collected during operational testing. Nonclassified SEF measures should be published and used as a project-control and management-control tool. The SEF would provide the FAA with a quantitative measure of the impact of security equipment and procedures.

Five-Year Deployment Plan

Decisions based on systems of systems analysis (e.g., TAAS) involve both management and cost factors, which are airport and airline specific. Stakeholder[3] involvement, therefore, will be crucial for the development of an effective deployment strategy. Furthermore, airline and airport buy-in will be critical to the successful implementation of the deployment strategy. The FAA did not provide the panel with a long-range (five-year) TAAS deployment plan developed jointly and agreed to by the FAA and other stakeholders. Thus, the panel concluded that the FAA has not obtained comprehensive airline buy-in for a long-term deployment plan that addresses all of the relevant issues, such as operator training, the optimal location of detection equipment, and the operational deployment of HULDs.

Recommendation. Within one year, in cooperation with the other stakeholders, the FAA should develop a five-year joint-deployment plan that includes cost, stakeholder responsibilities, quality measures, and other important factors. This plan should be a living document that is formally updated annually. Buy-in from all stakeholders will be necessary for the plan to be effective.

3. Assess the cost and advisability of requiring hardened cargo containers to enhance aviation security and reduce the required sensitivity of bomb-detection equipment.

Two HULDs (both LD-3 size) that conform to NAS-3610-2K2C airworthiness criterion have passed the FAA blast and shockholing[4] tests. The LD-3 container is used only on wide-body aircraft, however. Thus, no HULD concept for narrow-body aircraft has passed the FAA test, although 75 percent of the aircraft in service (as of 1994) are narrow-body aircraft, and more than 70 percent of bombing attempts have been against narrow-body aircraft.

The panel's greatest concern is that research on HULDs has not been conducted on a system-of-systems (SOS) basis and has not involved all of the stakeholders, mainly the airlines. So far, HULDs have largely been developed and designed as single stand-alone entities. Limited research has

[3] In this report the term *stakeholder* includes the FAA, the airlines, and the airports. Although there are certainly other stakeholders in aviation security, these three will have the most influence on the deployment strategy for aviation security equipment.

[4] A *shockholing* (or fragmentation) test measures the ability of a HULD to prevent perforation of its walls by a metal fragment traveling at a relatively high velocity.

been done on their role as part of a TAAS. Coordination with the airlines, airports, and aircraft manufacturers has been focused mainly on specific designs and utility requirements rather than on the interactions, boundary conditions, and trade-offs (including cost and operational considerations) of using HULDs along with other security measures, such as passenger profiling and baggage screening. The panel believes that alternative HULD designs may be more practical than existing designs in the TAAS context.

Recommendation. The FAA should continue to support research and development on hardened unit-loading devices (HULDs), including ongoing operational testing. If the FAA recommends, mandates, or regulates the use of HULDs, explosion-containment strategies for narrow-body aircraft, including the development of narrow-body HULDs and cargo-hold hardening concepts, should be investigated. However, the FAA should not deploy HULDs unless they are part of the TAAS joint five-year deployment plan.

4. On the basis of the assessments and determinations made under paragraphs (1), (2), and (3), identify the most promising technologies for improving the efficiency and cost effectiveness of weapons and explosives detection.

The data were not sufficient for a comprehensive assessment of available technologies for improving aviation security. Therefore, at this time the panel is not able to identify or recommend the most promising technologies for improving the efficiency and cost effectiveness of weapons and explosives detection. If the recommendations in this report are followed, these data will become available for subsequent assessments.

REFERENCES

DOT (U.S. Department of Transportation). 1998. Aviation Security: Federal Aviation Administration. Washington, D.C.: U.S. Department of Transportation, Office of the Inspector General. Also available on line: *http://www.dot.gov/oig/audits/av1998134.html*

FBI (Federal Bureau of Investigation). 1995. Terrorism in the United States in 1995. Available on line at: *http://www.fbi.gov/publish/terror/terrorin.htm*

GAO (General Accounting Office). 1998. Aviation Security: Implementation of Recommendations Is Under Way, But Completion Will Take Several Years. GAO/RCED-98-102. Washington, D.C.: General Accounting Office. Also available on line at: *http://www.gao.gov/AIndexFY98/abstracts/rc98102.htm*

White House Commission on Aviation Safety and Security. 1997. Final Report to the President. Also available on line at: *http://www.aviationcommission.dot.gov/*

Zuckerman, M.B. 1996. Are order and liberty at odds? U.S. News and World Report 121(5): 64.

1

Introduction

On April 22, 1995, FBI agents took custody of Abdul Hakim Murad from Philippine authorities. He had been arrested after a fire broke out in a Manila apartment where he, Ramzi Yousef, and another associate were living and where officials found explosives and bomb-making materials (FBI, 1995). The fire may well have prevented the worst terrorist attack against civil aviation in history. Yousef was later indicted for the 1994 bombing of Philippine Airlines Flight 434, which was determined to be a test run for a plot to blow up 11 American planes simultaneously (Zuckerman, 1996). Although it is horrifying to speculate on what might have happened if the fire had not broken out in Murad's apartment, it is more constructive to focus on what has been done—and what is being done—to improve aviation security.

Arguably the greatest progress in the last 30 years in the fight against terrorist attacks on aircraft has been made in the 10 years since the devastating bombing of Pan Am Flight 103 on December 21, 1988 (Figure 1-1). Although it is difficult to prove a cause-and-effect relationship between government action and the reduction in bombings, three laws passed by Congress (Box 1-1) have undoubtedly had an impact.

The three laws passed by Congress have facilitated the development and deployment of security equipment and procedures, which have improved aviation security. In 1997, the Federal Aviation Administration (FAA) was directed by President Clinton and authorized by Congress to deploy 54 FAA-certified explosives-detection systems[1] (EDSs) and more than 400 trace-detection systems in airports around the country. The FAA created the Security Equipment Integrated Product Team (SEIPT) to manage this deployment. The SEIPT assessed the availability of explosives-detection equipment and formulated a plan to deploy this equipment in airports throughout the United States. In a separate program, the FAA began testing hardened unit-loading devices (HULDs) designed to contain an explosive blast. Several HULDs are now undergoing operational testing by commercial air carriers.

Although substantial progress has been made, opportunities remain for the development and deployment of technologies that will make commercial aviation in the twenty-first century even safer. In the future, explosives-detection equipment must have higher throughput rates, lower false-alarm rates, and greater flexibility to detect different types of threat materials. HULDs must be proven to be airworthy and their tare (empty) weight reduced. Even if all of these challenges are met, these technologies must be deployed in a manner that provides maximum protection from terrorist attacks against commercial aircraft.

DEPLOYED TECHNOLOGIES

Aviation security equipment and procedures include the following: bulk[2] explosives-detection equipment, trace explosives-detection equipment, HULDs, computer-assisted passenger screening (CAPS), and positive-passenger bag matching (PPBM) (Table 1-1).

The congressionally mandated deployment of bulk explosives-detection equipment began with the installation of the first FAA-certified EDS (the InVision CTX-5000) and continued throughout 1998. The installation of trace explosives-detection equipment and the implementation of CAPS and PPBM were scheduled for the same time period. Two HULD designs (both LD-3 size) that conform to

[1] The following terminology is used throughout this report. An *explosives-detection system (EDS)* is a self-contained unit composed of one or more integrated devices that has passed the FAA's certification test. An *explosives-detection device* is an instrument that incorporates a single detection method to detect one or more categories of explosive material. *Explosives-detection equipment* is any equipment, certified or not, that can be used to detect explosives.

[2] In this report, the term *bulk explosives* includes all forms and configurations of explosives at threat level (e.g., shaped explosives, sheet explosives, etc.).

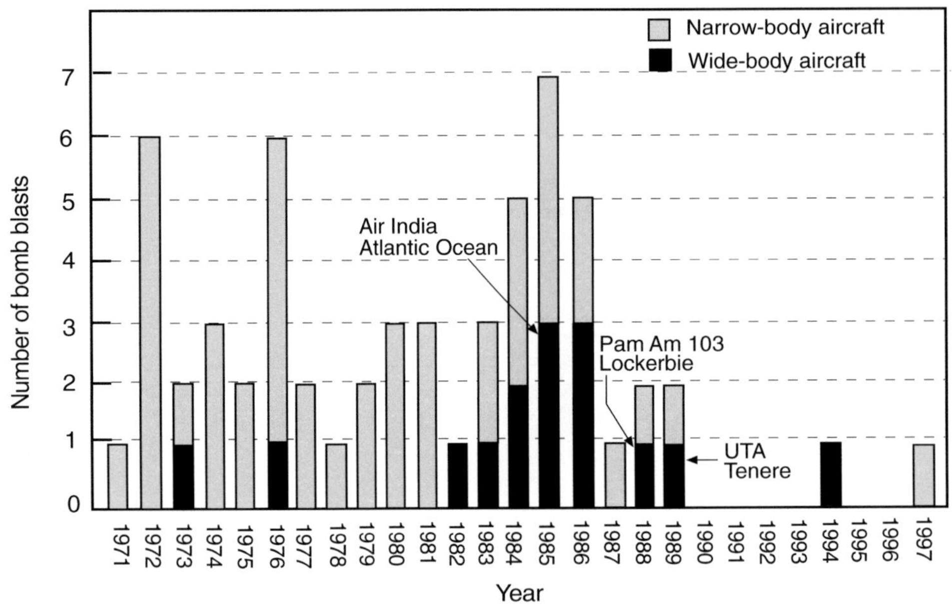

FIGURE 1-1 The distribution of aircraft bomb blasts between 1971 and 1997.

NAS-3610-2K2C airworthiness criterion have passed the FAA blast and shockholing[3] tests. Ten of these HULDs have been delivered to three different airlines for operational testing over the next year.

The FAA/SEIPT is behind schedule in the deployment of aviation security equipment (GAO, 1998; DOT, 1998). When Congress provided $144.2 million for the purchase of commercially available security-screening equipment, the FAA/SEIPT planned to deploy 54 certified EDSs and 489 trace-detection devices by December 1997 (GAO, 1998). The FAA also planned to have CAPS fully implemented by December 1997. When it became clear that these goals could not be met, the FAA set a new goal of deploying 54 certified EDSs, 22 noncertified bulk explosives-detection devices, and 489 trace explosives-detection devices by December 31, 1998, and of implementing CAPS by November 1998. As of January 1, 1999, more than 70 certified explosives-detection systems, six noncertified bulk explosives detection devices, and 366 trace-detection devices had been installed in airports.

TOTAL ARCHITECTURE FOR AVIATION SECURITY

Protecting civil aviation against terrorist threats is a complex systems problem that has no perfect solution. Significant compromises have to be made in security systems to achieve the highest level of security at an affordable cost while at the same time maintaining the efficiency of air travel. Improvements in aviation security can best be quantified by a security enhancement factor (SEF) that measures improvements in security compared to a baseline level of security in a given year. However, SEF is an exceedingly complex measure because the threats to aviation security, and the available security technologies, are variable and time dependent. In fact, many different detection and protection techniques are being used to counter several different threats, which in turn are influenced by many factors, including geographic location, weather conditions, and the political climate.

The U.S. Department of Defense has been faced with similarly complex systems problems and, through experience, has come to address them in a system-of-systems (SOS) framework. An SOS is a complex of systems, each of which is characterized by measures of performance against threats and costs for acquisition and deployment. The SOS concept can be used to optimize a complex system by providing a top-level perspective. For example, instead of optimizing a particular system *A* for performance and cost, the optimization of the performance and cost of the SOS as a whole (which might consist of systems *A, B, C,* and *D*) may require that performance requirements for system *A* be reduced or even that system *A* be eliminated.

An SOS concept would enable FAA management to mount a layered defense against a dynamic threat. A well

[3] A *shockholing* (or fragmentation) test measures the ability of a HULD to prevent perforation of its walls by a metal fragment traveling at a relatively high velocity.

BOX 1-1 Public Laws on Aviation Security since 1988

Public Law 101-604. After the bombing of Pan Am Flight 103, a Commission on Airline Security and Terrorism was created by President Bush, which led to his signing of the Aviation Security Improvement Act of 1990 (Public Law 101-604). This act directed the FAA to accelerate and expand the research, development, and implementation of technologies and procedures to counteract terrorist acts against civil aviation; determine the amounts and types of explosives that could cause catastrophic damage to an airplane; and established the position of assistant administrator for civil aviation security.

Public Law 104-264. In 1996, President Clinton established the White House Commission on Aviation Safety and Security. The recommendations of this commission led to a directive in the Federal Aviation Reauthorization Act of 1996 instructing the FAA to deploy certified and noncertified explosives-detection equipment.

Public Law 104-208. The Omnibus Consolidated Appropriations Act of 1997 provided $144.2 million for the FAA to purchase and assist in the installation of advanced security equipment. This act was a significant departure from previous policy, under which air carriers were responsible for purchasing and deploying aviation security equipment.

TABLE 1-1 Selected Aviation Security Equipment and Procedures

Technology	Description
Computer-assisted passenger screening (CAPS)	CAPS is a system that utilizes a passenger's reservation record to determine whether the passenger can be removed from consideration as a potential threat. If the passenger cannot be cleared (i.e., determined not to be a threat), CAPS prompts the check-in agent to request additional information from the passenger for further review. If this information is still insufficient to clear the passenger, the passenger's bags and the passenger are considered "selectees" and are routed through additional security procedures.
Positive passenger-bag match (PPBM)	PPBM is a security procedure that matches the passenger's checked baggage with the passenger to ensure that baggage is not loaded aboard an airplane unless the passenger also boards. This security measure is implemented for all outbound international flights and for some domestic flights.
FAA-certified explosives-detection systems (EDS)	An EDS is a self-contained unit composed of one or more integrated explosives-detection devices that have passed the FAA's certification test. As of April 1999 only computed-tomography-based technologies have passed the FAA bulk explosives-detection certification tests (e.g., InVision CTX-5000, CTX-5000 SP, and CTX-5500 DS).
Bulk explosives-detection equipment	Bulk explosives-detection equipment includes any explosives-detection device or system that remotely senses some physical or chemical property of an object under investigation to determine if it is an explosive. This equipment, primarily used for checked baggage, consists of quadrupole resonance and advanced x-ray technologies, including radiography and tomography.
Trace explosives-detection devices (TEDDs)	TEDDs involve the collection of particles or vapor from the object under investigation to determine if an explosive is present. TEDDs are being deployed for several threat vectors: carry-on baggage (especially electronic devices), passengers, checked baggage, and cargo. TEDDs employ a variety of techniques for detecting vapors, particles, or both, which include chemiluminescence, ion mobility spectroscopy, and gas chromatography. TEDDs do not indicate the amount of explosive present and hence do not reveal the presence of a bomb, except inferentially.
Hardened unit-loading devices (HULDs)	A HULD is a specially designed baggage container that can contain the effects of an internal explosion without causing damage to the aircraft. A design by Galaxy Scientific passed the FAA blast test in March 1998. A second Galaxy Scientific design passed the FAA blast test in January 1999. To study operational performance and reliability, the FAA deployed 10 Galaxy HULDs in 1999.

defined SOS framework enhances communication among interested parties, even if the analysis is only semi-quantitative. A well understood measure (e.g., SEF) of the efficacy of the SOS would also provide a credible basis for allocating budgets for system improvements. The SOS approach would enable the FAA to describe and assess the deployment of explosives-detection equipment, HULDs, CAPS, and PPBM, as well as other security equipment and procedures—including the performance of human operators. Equipment, procedures, and human operators work hand in glove with other units in the overall airport security system and should be measured and assessed in that framework. In the panel's opinion the only way to assess an intertwined system with feedback or feed-forward control loops is through an SOS approach. Therefore, the panel adopted an SOS approach to devise a total architecture for aviation security (TAAS) as a framework for assessing aviation security.

REPORT ORGANIZATION

This report presents an SOS approach to assessing aviation security, introduces an SEF, and describes the FAA's progress in deploying aviation security equipment and procedures. Recommendations are also made for future deployments of security equipment and implementations of security procedures. The TAAS and SEF are introduced and described in detail in Chapter 2. Chapter 3 defines the roles and responsibilities of the FAA, air carriers, airports, and independent security contractors in the deployment and maintenance of the performance of security equipment and procedures. Chapter 3 also describes a management framework for the deployment. Cargo and baggage handling are discussed in Chapter 4, providing a context for the implementation of security equipment and procedures described in Chapters 5 through 8.

Explosion-resistant containers, or HULDs, are described in Chapter 5. In Chapter 6, the FAA's progress in the mandated deployment of bulk explosives-detection equipment is described, as well as the results of performance testing, including detection rates, false-alarm rates, and throughput rates. The FAA's progress in deploying trace explosives-detection equipment is discussed in Chapter 7, which also includes the panel's rationale for recommending that tests be developed to measure the performance of these devices. CAPS and PPBM are discussed in Chapter 8, including a timeline for their deployment and a description of how they can be used together. Chapter 9 is a discussion of human factors in the operation of security equipment. Evaluations of airport architectures and their relationships to TAAS are presented in Chapter 10. The panel's overarching high-level conclusions and recommendations are presented in Chapter 11.

2

Grand Architecture

Protecting commercial aviation from terrorist threats is a complex systems problem. As illustrated in Figure 2-1, there are many paths by which a threat material (i.e., a bomb) can endanger the security of an aircraft. These threat vectors include checked baggage, cargo, mail, passengers and their carry-on bags, flight crews, catering and service personnel, and missiles. Aviation security must protect aircraft against attack—from explosives, chemical agents, biological agents, and other threat items—by all of these threat vectors, and possibly others. There is no perfect defense against all threats to commercial aviation, and optimizing aviation security with respect to performance, cost, and efficiency of air travel will ultimately require compromises in the selection of security equipment, procedures, and personnel.

The focus of this study is on the security measures deployed by the FAA for detecting and containing explosives introduced by two threat vectors, checked baggage and carry-on baggage. Although this does not encompass all potential threat vectors, designing and implementing effective security measures to address even these two vectors together are a complex systems problem. Because no single "silver bullet" technology can protect against all threat vectors, the

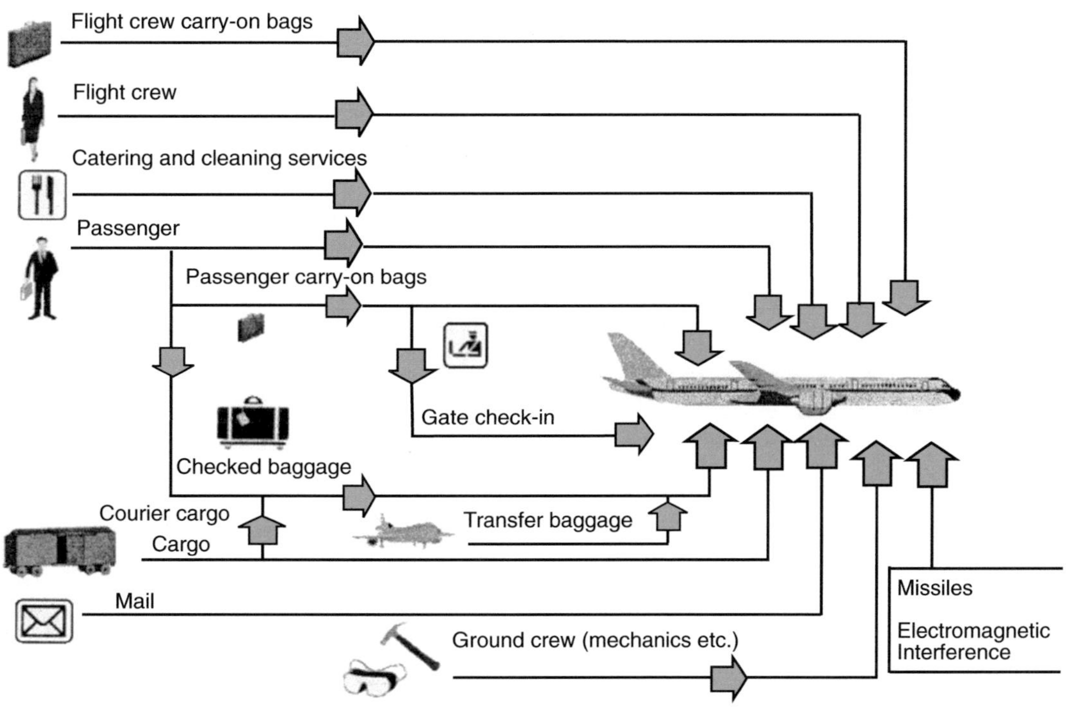

FIGURE 2-1 Threat vectors. The paths by which people, baggage, and equipment board a plane are also routes by which threats may board a plane.

overall security system must be a multilayered network of subsystems. The performance of the overall system depends on the technical capabilities of the detection and containment equipment and the performance of the equipment operators, as well as a host of environmental, management, and policy factors. System performance is also affected by operator training, maintenance practices, management priorities and loyalties, and the nature of the threat itself. Like all of the other factors that influence system performance, the nature of the threat is variable and time dependent, which contributes to the complexity of the overall security system.

Complex systems problems can best be understood in an SOS framework. The primary performance measure for an SOS for aviation security is, of course, protection against explosives. Performance can best be described as an SEF, which measures the improvement in security relative to a specified baseline (e.g., the security afforded by the system configuration in a particular year). SEF is a simple measure that condenses an exceedingly complex SOS into a single parameter that can accommodate the variabilities and time dependencies of protective measures, as well as of the threat itself. The SEF compares an improved security system to the previous system in terms of the probability that a bomb will be taken aboard a plane (and cause catastrophic damage to it) by comparing the number of test bombs that had defeated an older security system to the number of test bombs that have defeated the newer security system. The panel concluded that the SEF is one way to reduce the complexity of analyzing the aviation security system.

In the context of a well defined SOS framework, security systems can be designed or modified to optimize aviation security. As the responsibility for security measures becomes increasingly diffuse and more and more liability claims are being disputed, the need for an SOS framework for interpreting the viability of security systems is becoming more urgent. Although predicting the performance of an aviation security system against a terrorist event is difficult, an SOS approach makes it possible to estimate the performance range of a security system based on thorough and realistic operational testing.

For the reasons cited above, the panel adopted an SOS approach to assess the FAA's deployment of explosives-detection equipment, HULDs, and security procedures. Security equipment must work in concert with other units in the overall airport security system and, therefore, should be measured and assessed in this larger context. The panel developed the TAAS (total architecture for aviation security) as a framework for assessing their performance.[1] The panel presents a rationale for using a high-level systems approach and a methodology for continuously monitoring and upgrading the TAAS in response to changes in the threat environment, security technologies, and procedures. In addition, qualitative requirements for the overall system architecture are related to performance and operational characteristics of the subsystem components, including policies. Once these concepts are endorsed by the FAA, the airlines, and airports quantitative measures can be developed to assess and optimize improvements to the TAAS.

TOTAL ARCHITECTURE FOR AVIATION SECURITY CONCEPTS

Figure 2-2 illustrates a top-level TAAS for explosives detection and containment. The first step in the analysis is the description of the real-world threat. This threat is variable and can be described as a range of probabilities that a specified amount, type, and configuration of explosive will enter the system, or the amount of explosive can be a random variable whose distribution varies with the type of explosive. Articles accompanying the explosive and other features of the operating environment (e.g., passenger characteristics, air traffic) also appear as random variables. Because there are many ways to define the explosives threat, the modeling of the threat environment and its dependencies on the venue and other factors will have to be carefully considered.

The security layers in Figure 2-2 are shown as generic components labeled *A, B, C,* etc. These components include physical security,[2] metal detection, bulk and trace explosives-detection equipment, operator inspection of x-ray screens for carry-on luggage, CAPS, PPBM, and the possible containment of baggage in HULDs. Each component, including associated operational protocols (training, calibration, maintenance, and other procedures), comprises a subsystem in the TAAS. Each subsystem is described by measures for determining its effectiveness, such as throughput rate, false-alarm rate, operational cost, installation cost, and probability of detection. Links between the subsystems are an important aspect of the TAAS. For instance, passenger-profiling information (from CAPS) could be provided to x-ray screening consoles to enable operators to rapidly resolve a security alert from a particular bag (an example of feed forward). Because these links affect the top-level performance of the TAAS, the flow pattern is also an important part of the analysis. Two routes through the TAAS are shown in Figure 2-2 (*A-B-C* and *D-E-F-G*) to indicate that there may be more than one way through the TAAS to the plane.

The role of various authorities in the management of the TAAS may add to its complexity. Currently, airlines have significant discretion over the deployment of aviation security subsystems. Baggage, cargo, and mail flow vary among

[1] In a recent report, *Aviation Security Technology Integration Plan,* the FAA uses a similar systems-architecture perspective for planning aviation security strategies (LaMonica et al., 1995).

[2] *Physical security* includes measures to control access to concourses, gates, and airplanes to ensure that passengers and bags go through the security system.

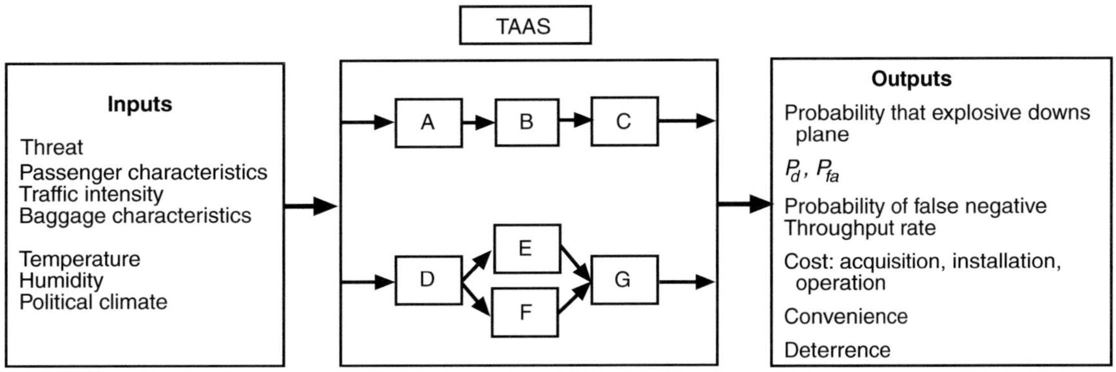

FIGURE 2-2 A top-level total architecture for aviation security (TAAS).

domestic and foreign air carriers, change with airline transfer protocols, and change again with the imposition of regulatory policies by various agencies, such as the FAA and local airport authorities. Because of the large variability in baggage, cargo, and mail flow, the TAAS for each airport will be specific to that facility. Thus, an evaluation of a deployed system will necessarily be specific to a particular site, and a top-level strategy for future deployments must be based on both universal and site-specific characteristics.

Simplifying, comparing, and analyzing various TAAS configurations will require an overall SEF. Because of the variable nature of the threat and other components and the probabilistic behavior of the subsystems, the TAAS performance measures can also be described as random quantities. A primary output of the assessment of a particular TAAS would be the probability that an explosive threat will bring down an aircraft, which is expressed as a probability distribution over the various threat amounts. One measure of the performance of the TAAS will be the SEF, which measures the reduction, relative to the baseline architecture, in the probability that a threat will bring down an aircraft.

Improved security is the ultimate goal but not the only performance measure of the TAAS. Every system architecture has a number of costs and customer (passenger) convenience features that must be traded-off against the SEF. For example, airline passengers are not likely to tolerate extra delays for extensive scrutiny of their baggage in peacetime when no threat is perceived. Similarly, substantial purchase and deployment costs by the FAA would not be tenable for small improvements in security. Additional costs to the airlines for the TAAS would surely meet with customer resistance if the cost was translated to significantly higher ticket prices. These and other performance measures are also shown as outputs in Figure 2-2. One way to assess security enhancement/cost trade-offs would be to determine the cost per passenger per percent of increase in the probability of detection of an explosive threat (Hammar, 1998). An evaluation of security architectures and additional discussion of the TAAS concept are contained in Chapter 10.

TOTAL ARCHITECTURE FOR AVIATION SECURITY SUBSYSTEMS

The explosives-detection equipment and containment equipment are the security subsystems comprising the networks in the TAAS in Figure 2-2. Figure 2-3 shows the subsystems that were in place prior to the 1997–1998 deployment, including conventional x-ray radiography for scanning carry-on baggage, metal detector portals for screening passengers, canine teams for screening checked bags, physical searches of baggage, and limited passenger-bag matching. A network like the one in Figure 2-3 could be used as the baseline architecture for the SEF.

In response to the congressional mandate (PL 104-264), the FAA focused on the development and deployment of the following security measures: CAPS, FAA-certified EDSs, noncertified bulk explosives-detection equipment, TEDDs, PPBM, and HULDs. Current airport security systems also include conventional x-ray scanning, physical searches, metal detectors, and canine teams. Figure 2-4 shows a representative TAAS in the early stages of the FAA's mandated deployment. To evaluate the effectiveness of this deployment, one must estimate the improvement in security afforded by the TAAS in Figure 2-4 over that in Figure 2-3 (including the performance of detection and containment equipment, management policies, and human factors).

Once reliable and statistically significant data are available on the performance characteristics of individual subsystems (or security measures) and their dependencies, the optimal configuration of the TAAS can be determined. In general, complementary detection devices yield higher SEFs than redundant identical devices. The TAAS provides a basis for a systematic analysis of trade-offs among subsystems.

SECURITY ENHANCEMENT

The purpose of a system that includes hardened containers would be to reduce the probability that an onboard explosion would bring down a plane. Because hardened containers

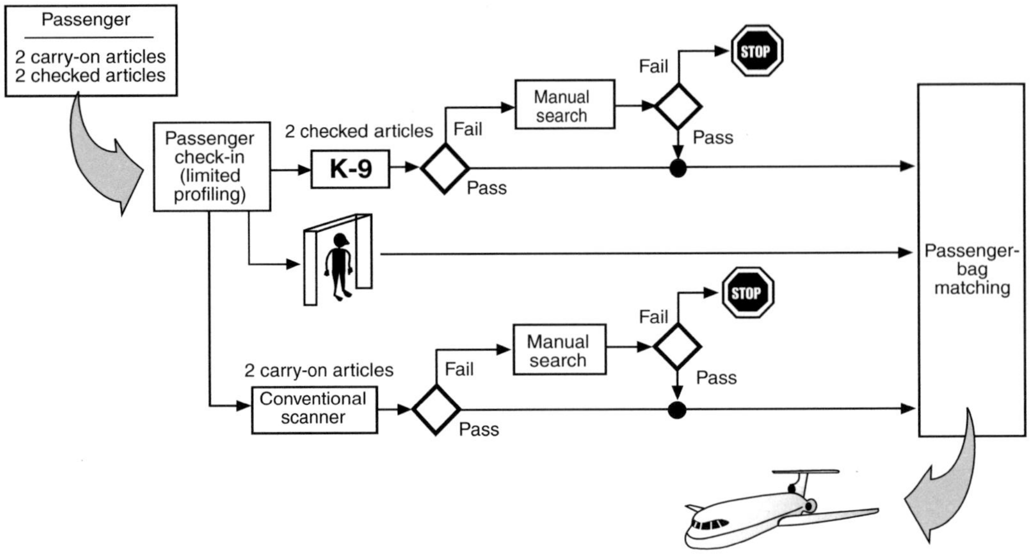

FIGURE 2-3 Notional airport security configuration for international flights prior to the 1997–1998 deployment. Source: Dickey and Fuqua (1998).

have not yet been deployed, security enhancement is currently evaluated by the decrease in the probability that a bomb would be taken onto a plane. Thus the critical system-performance measure is the proportion of bombs that defeat the security system. To evaluate the probability that a bomb would defeat a security system, realistic bomb simulants (e.g., the modular test set) must be used for blind operational tests (e.g., so-called red-team testing). The SEF of a System *B* relative to a baseline system (System *A*) is defined as:

$$\text{SEF} = \frac{\textit{proportion of test bombs that defeat System } \text{A}}{\textit{proportion of test bombs that defeat System } \text{B}}$$

The same set of test bombs and the same testing procedures are used for both systems. The probability that a bomb would defeat a security system is conditional on a bomb being present and not on the probability of bombing attempts, which will vary from the baseline year.

System *A* will probably change over time. Therefore, the

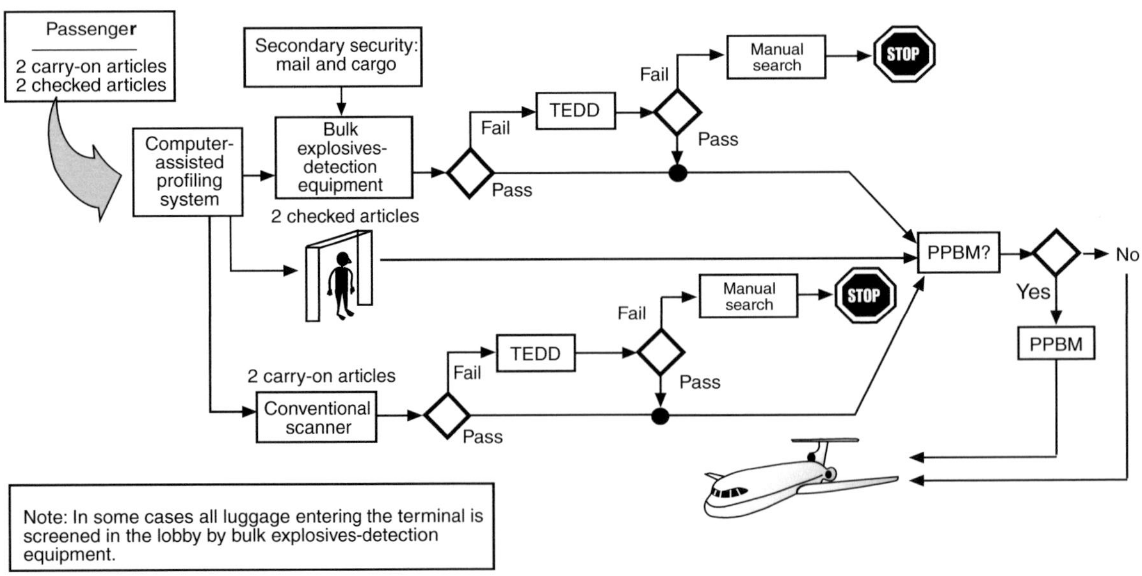

FIGURE 2-4 Notional aviation security configuration for international flights during the early stages of the 1997–1998 deployment. Source: Dickey and Fuqua (1998).

baseline system will have to be redefined as new and improved equipment becomes available and as the threat evolves with time. For example, if a new explosive compound that is difficult to detect with deployed equipment suddenly becomes prevalent among terrorists, the SEF would actually decrease. In this case, the security system that was in place prior to the appearance of the new explosive compound would not be relevant as a baseline. The "new" baseline system (System *A*) would be the system in place at the time the new explosive compound became known. An improved system (System *B*) would be the security system after a change has been made to System *A* to improve performance. Thus, the SEF would be the ratio of simulated explosives (including a simulant of the new explosive compound) that defeated System *B* to the simulated explosives that defeated System *A*. This scenario also has implications for determining the SEF when new equipment is deployed to an existing TAAS.

ANALYSIS TECHNIQUES

Monte Carlo techniques can be used to calculate the distributions for the outputs shown in Figure 2-2. The distributions are based on repeated inputs (from TAAS performance measurements for selected component and threat distributions). Because of the randomness of the system, different Monte Carlo runs of the same input values may yield different outputs. Monte Carlo methods, which were developed to assess the susceptibility or risk of complex systems to various threats (or failures), are often called probabilistic risk assessments (PRAs) (Aven, 1992; Henly and Kumamoto, 1992). Three aspects of PRAs that are specific to the TAAS analysis are listed below.

1. The results of PRAs are heavily dependent on the underlying assumptions of the distributions of the input variables and the performance characteristics of the layered security subsystems. All subsystems must, therefore, be carefully evaluated in field operation and all dependencies investigated. The results of PRAs will only be as valid as the probabilistic models used to specify the TAAS.
2. A PRA of the TAAS using performance measures for actual bombing attempts would result in a very small input distribution. When this occurs, obtaining a precise output distribution (e.g., the probability that an explosive will bring down a plane) requires a very large number of Monte Carlo runs. For this reason, realistic simulated bomb attacks[3] are the only reliable way to test the capability of the TAAS. An SEF measure that compares actual successful attacks against the baseline and the improved TAAS would yield a ratio of two very small quantities, making it difficult to estimate improvements. Rather than evaluating the real threat of attack, the SEF shows the improvement in the system once a threat is introduced. This approach focuses on the critical false-negative rate rather than the false-alarm rate.
3. It may be possible to simplify the threat distribution, and therefore the PRA, by establishing the likelihood that a particular explosive will be used. For example if terrorists tend to use only one type of explosive, (e.g., explosive *X*) 99 percent of the time and explosive *Y* only 1 percent of the time, only explosive *X* would have to be considered in the analysis. Assume the distribution of the amount (Q) of explosive is sufficient to produce a distribution of possible losses for different planes without HULDs and another distribution of an amount greater than Q (Q+) is required for planes with HULDs, and assume that there is a 90 percent chance that an amount Q will bring down a plane without a HULD and a 90 percent chance that an amount Q+ will bring down a plane with a HULD. Then Q or Q+ for a specific explosive can be used as simple parameters to evaluate the success of the TAAS without requiring the evaluation of every distribution.

THE FAA'S DEPLOYMENT STRATEGY

The FAA's deployment of advanced explosives-detection technologies was initiated with an allocation of funds by Congress specifically for this purpose (PL 104-208). This equipment had to be deployed on an accelerated schedule because of time constraints on the funds (PL 104-264, PL 104-208), which precluded the development of a strategic deployment plan. Even though many deployments were less than optimal and some airport surveys were not completed, the placement of advanced technologies into the field has provided valuable data for future deployments (e.g., effect of location of the explosives-detection equipment and the flight destination on alarm rates). A great deal has also been learned about the deployment process, particularly the fundamental importance of securing the cooperation of airport authorities, the airlines, and FAA personnel.

Because the FAA was aware of the importance of these field data, steps were taken to ensure that the data were analyzed and stored (Dickey, 1998; Fuqua and O'Brien, 1998). Based on these data, the FAA now has the opportunity and means of pursuing a genuine deployment strategy. Indeed, the panel observed that the FAA has taken the following steps toward comprehensive strategic deployment:

- The FAA's Office of Policy and Planning for Civil Aviation Security has developed a series of potential scenarios (dubbed "end states") for future aviation se-

[3] A simulated attack implies that the dynamic range of the testing process is sufficient to evaluate the differences between the baseline and improved TAAS.

curity and is developing requirements for security checkpoints; the training and performance of personnel; and the handling of cargo, mail, and checked baggage. In addition, this office envisions airport-airline-FAA partnerships to facilitate the deployment of security equipment and procedures and provide assistance in risk management, vulnerability assessments, and contingency planning. As part of the strategic planning, the office has made a cost comparison of the widespread deployment of certified EDSs and the deployment of a slower, cheaper (i.e., noncertified) alternative (Fainberg, 1998).

- The Aviation Security Technology Integration Plan uses a systems architecture for planning aviation security strategies. The FAA has developed a comprehensive, although qualitative, plan for continuously monitoring and improving this architecture (LaMonica et al., 1995; Polillo, 1998).
- The SEIPT Operational Assessment Report describes data requirements for planning and fielding security equipment (Fuqua and O'Brien, 1998). The FAA's system-assessment concept outlines a plan for obtaining operational data. A deployment-analysis database is also being planned for obtaining and maintaining operational data that will be readily available for future assessments and planning (Hammar, 1998; Fuqua, 1998).
- The Passenger Bag Flow Model is a large simulation model of the flow of passengers and baggage in an airport being developed by the FAA. The model will incorporate operational performance data and airport characteristics, such as layout, flight schedules, and passenger base, to simulate overall system behavior (Hammar, 1998).

CONCLUSIONS AND RECOMMENDATIONS

The FAA is now in a position to adopt a systems approach for the strategic deployment of security equipment and procedures. This approach will require active participation by airlines and airports, a systematic process for collecting operational data, and a well defined SEF-type measure to reduce the complexity of the analysis. An SOS framework, such as the TAAS described above, would be capable of describing and assessing the deployment of explosives-detection equipment and other security measures.

Recommendation. The FAA should define a total architecture for aviation security (TAAS) for describing and assessing the deployment of explosives-detection equipment, hardened unit-loading devices, and security procedures.

Recommendation. The FAA should formulate a security enhancement factor (SEF) for the integrated total architecture for aviation security based on data collected during blind operational testing.

Recommendation. The FAA should aggressively define operational performance metrics for security subsystems and for the total architecture for aviation security as a whole. The FAA should establish an action team whose principal task is the systematic collection of operational data. These data should be systematically placed into a database and made available to those who have a "need to know."

3

Roles and Responsibilities

Many factors affect the performance and effectiveness of security systems, including physical location, operator training, and the presence of law enforcement personnel. Although the FAA has the overall responsibility for the effectiveness of aviation security, the buy-in of other stakeholders is critical for funding security equipment and for implementing a long-term security strategy, such as the TAAS (total architecture for aviation security) described in Chapter 2. In this chapter, the roles and responsibilities of the FAA, the air carriers, airports, and independent security contractors are discussed (see Figure 3-1).

The U.S. civil aviation security program today is a combination of laws, regulations, and resources for protecting the industry and the traveling public against terrorism and other criminal acts. The program is a system of shared and complementary responsibilities involving the federal government, airport operators, air carriers, and passengers. The FAA sets standards and guidelines, and airports and air carriers implement them. Airline passengers and the users of air cargo, who are the ultimate beneficiaries of the program, pay for the program through security surcharges included in the prices of airline tickets and cargo shipments. Although

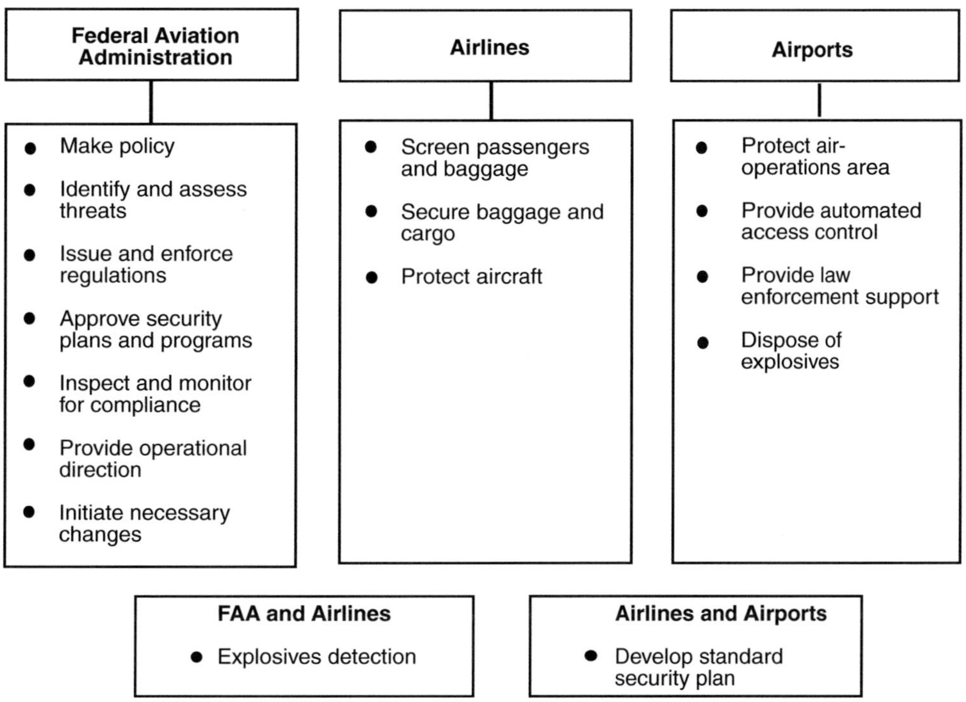

FIGURE 3-1 Responsibilities for civil aviation security.

authority can be delegated or shared (e.g., a private security contractor might operate explosives-detection equipment), the ultimate responsibility for the safety and security of civil aviation rests with the state, in this case the FAA.

FEDERAL AVIATION ADMINISTRATION

The FAA has existed in various forms and under various names since 1926 when Congress passed the first of many federal aviation laws. The Federal Aviation Act of 1958 created an independent Federal Aviation Agency with the responsibility of establishing air safety regulations and certification requirements. In 1967, the agency was renamed the Federal Aviation Administration and transferred to the newly created U.S. Department of Transportation.

FAA headquarters is located in Washington, D.C. Nine regional and three international offices administer the field elements located in their geographical areas of responsibility. The FAA administrator makes policies, issues regulations, and provides overall direction for the FAA's functional programs. Field elements perform operational functions and enforce aviation regulations. The associate administrator for civil aviation security is responsible for assessing and addressing the threat to civil aviation security.

Aviation Security Research and Development Program

The mission of the Aviation Security Research and Development Program is to generate and disseminate expertise and knowledge in technologies relevant to the security of the civil aviation industry and to provide technical support services for the associate administrator for civil aviation security. All agency research and development projects are consolidated into the Office of Aviation Research headed by the associate administrator for research and acquisitions.

Deployment of Equipment

In 1997, Congress mandated that the FAA deploy commercially available explosives-detection equipment in U.S. airports. Given the nature of the government mandate, the FAA was the logical organization to evaluate, certify, and deploy the initial explosives-detection equipment. The urgency of placing security equipment in U.S. airports, combined with the congressional mandate that the FAA administer the process, placed the FAA in the unique position of being both project manager and systems integrator.

The FAA administrator, through the headquarters staff, initially deployed explosives-detection equipment to selected air carriers serving the Atlanta and San Francisco airports; subsequently equipment was deployed to air carriers at other large U.S. airports. The FAA worked closely with the manufacturers of explosives-detection equipment to select, test, and install equipment and provide operator training for the new hardware. The FAA had the primary responsibility for all aspects of the initial deployment. Other stakeholders, including the installation-site air carriers and airport operators, played only consulting and supporting roles.

AIRPORTS

More than 25 years ago, in 1971, the FAA issued Federal Aviation Regulation (FAR) Part 107, which gave airport operators the responsibility of protecting against unauthorized access to the air-operation areas of airports. Numerous changes have been made to this FAR over time to keep pace with changing security needs. Basically, FAR Part 107 prescribes aviation security rules governing the operation of each airport that regularly serves scheduled passenger operations. FAA security rules apply to 458 of the 667 certified U.S. airports, which are divided into six categories (Table 3-1).

The airport operator is the administrator and manager of an airport that regularly serves scheduled passengers of a certificate holder (air carrier) or a foreign air carrier subject to the security program described in FAR Part 108. Ownership and operation of domestic airports vary considerably; airports may be publicly or privately owned; they can be operated by a city, a county, a state, or a specialized airport authority. For example, the New York Port Authority, a bistate commission, owns and operates John F. Kennedy International, LaGuardia, and Newark International airports; in Chicago, the largest commercial airports are city owned and operated; Baltimore-Washington International Airport is state owned; McCarren International Airport in Las Vegas is county owned.

Airport ownership also determines the law enforcement structure at an airport. The primary organizations providing this support service are state and local police forces and special airport-authority police. Regardless of the organization providing the law enforcement, the FAA requires that certain criteria be met to ensure a consistent level of service. Most airport operators also employ security forces for physical security in the airport. In some cases, this service is provided by private contractors.

Physical security at many airports is further subdivided, by exclusive-use agreements, between the airport operator and the air carriers serving the airport. Exclusive-use agreements transfer to the air carrier the responsibility for physical security in operational areas leased from the airport, including air-operations areas, cargo buildings, and airline spaces in the terminal buildings.

Airport operators are conscientious about their security responsibilities and knowledgeable of the specific security needs of their airports. Security can always be tighter, of course, but many trade-offs must be made to maintain adequate security without overly restricting the efficient movement of passengers through airports. Turning airports into armed camps with military-style security would certainly

TABLE 3-1 Airport Categories in the United States

Category	Description	Number
X	Largest and most complex airports according to any one of three criteria: • 25 million or more persons screened annually • 1 million or more international enplanements annually • special considerations (e.g., serves a large political community or faces possible threats based on local conditions)	19
I	Screens between 2 million and 25 million persons annually	58
II	Screens between 500,000 and 2 million persons annually	52
III	Screens fewer than 500,000 persons annually and serves aircraft with seating configurations for 61 or more passengers	137
IV	Serves operations where screening is performed and passengers are deplaned into a sterile area of the airport or that screen pursuant to company policies and serves aircraft with 60 seats or less	168
V	Serves aircraft that seat more than 30 but less than 61 passengers where screening is not conducted and passengers are not deplaned into a sterile area	24

decrease their vulnerability to criminal or terrorist acts, but the economic and social effects could be detrimental.

During normal airport operations, the level of security is commensurate with the perceived threat and the movement of people, cargo, vehicles, and aircraft. To minimize disruptions, airport operators rely on contingency plans to upgrade security quickly in response to new intelligence or in emergency situations. Airport operators, air carriers, and other airport tenants must be capable of reacting immediately, and in a coordinated way, to an increased threat that requires a higher level of security.

Airport Security Programs

Airports rely on exclusively developed security programs that are approved by the FAA. These programs are primarily designed to provide a secure environment for airplane operations, control the movement of people and ground vehicles, and prevent unauthorized access to the air-operation areas. The plan must also include measures for protecting both air-operation areas and the publicly accessible land side (i.e., close-in public parking and terminal roadways) of the airport.

The terminal building presents unique security problems because public areas, restricted areas, and air-operation areas must be kept separate. Ultimately, the security plan for the terminal and ramps must allow passengers access to unrestricted areas while keeping unauthorized individuals from gaining access to restricted areas.

Security of Parked Aircraft

Airport operators must have facilities and procedures to prevent or deter persons and vehicles from gaining unauthorized access to air-operation areas. Air carriers are required to prohibit unauthorized access to their aircraft and to conduct a security inspection of an airplane if it has been left unattended and before it is placed in service.

Law Enforcement Support

The presence of law enforcement officers (e.g., police officers) is required at airports. In addition, most airlines contract guard agencies to provide personnel to perform predeparture screening. Both contract security guards and law enforcement officers may be stationed at screening checkpoints.

Official law enforcement officers must be authorized to carry and use firearms, vested with arrest authority, readily identifiable by uniform and badge or other indicia of authority, and must have completed a training program defined in the airport security program. There are almost as many approaches to satisfying these requirements as there are airports. Providing law enforcement officers is the responsibility of the airport operator, not the air carrier. Even though the cost of law enforcement is borne indirectly by the carriers, they have no power over whether airports employ city or state police officers or specially empowered guards.

Airport Consortia

In response to a recommendation by the White House Commission on Aviation Safety and Security (1997), the FAA directed parties responsible for aviation security to form consortia at 41 major U.S. airports. By mid-December 1997, 39 consortia had completed vulnerability assessments and developed action plans incorporating FAA-recommended procedural changes and requirements for advanced security technology. Voluntary security consortia are planned for more than 250 additional airports.

AIR CARRIERS

Air carriers are responsible for maintaining the security of passengers, baggage, and cargo entering the airplane in accordance with the FAA's standards and guidelines. FAR Part 108 requires that each U.S. air carrier adopt and carry out a security program for scheduled and public charter-passenger operations. The sole purpose of the air-carrier security program is to protect the traveling public from hijacking, sabotage, and other criminal acts. The FAA must approve all air-carrier security programs.

FAR Part 108 requires that each certificate holder designate a ground-security coordinator and an in-flight security coordinator to carry out the duties specified in the approved security program for each international and domestic flight. The pilot in command of each flight serves as the in-flight security coordinator.

Organizational Structure of U.S. Air Carriers

The organizational placement of the security function varies with the corporate structure of the airline. Among U.S. air carriers, security is a staff function with a reporting relationship to senior airline officials generally below the level of chief executive officer. Because security requirements can have a substantial impact on operations—including flight schedules and passenger processing times—security personnel work very closely with airline-operations officials. The responsibilities of an airline security office include interpreting FAA security regulations, setting policies and procedures for compliance by the airline, auditing and inspecting security operations, and representing the carrier in security-related matters. The airline security officer is also responsible for other security matters, such as theft and fraud.

The air carrier's local station manager is typically responsible for all operational activities at an individual airport and the day-to-day activities of the security contractor. Airline security at domestic airports is typically contracted out to private firms that provide trained personnel to operate the passenger screening checkpoints. Because of high costs, the air carriers encourage competition among security firms through a bidding process. However, even the lowest bidder must meet the minimum requirements. In cases where several airlines use the same security checkpoint at a concourse (as opposed to at the gate), one airline manages the checkpoint. The costs of operating and managing the checkpoint, however, are shared among the air carriers.

Predeparture Screening

The most visible aspect of domestic airline security is the screening of passengers and carry-on baggage, which was mandated by the FAA almost 30 years ago. The primary purpose of the procedures for domestic flights was to deter hijackers. Approximately 15,000 preboarding passenger screeners work in the United States for domestic and foreign air carriers. In 1995, they screened approximately 1.3 billion persons[1] at some 700 screening checkpoints. Three general arrangements of screening facilities at airports have been developed: the sterile concourse, the sterile boarding area, and departure-gate screening. The sterile concourse approach has proven to be the most attractive because it controls access by the inspection of persons and property at selected choke points and represents an exceptional cost savings for both air carriers and the airport. Instead of having to bear the cost of a workforce sufficient to search passengers at each gate and posting law enforcement officers at each gate, a central screening point at the entrance to a concourse can serve all of the gates on the concourse. Central screening checkpoints include x-ray machines, metal detectors, and security personnel.

The next best alternative for predeparture screening is the sterile boarding or holding area. In this arrangement, a sterile area is created at the flight check-in point, usually by securing the boarding lounge from the concourse or other adjacent terminal areas, to isolate passengers who have been screened from physical contact with unscreened persons.

The third, and least desirable, arrangement is departure-gate screening. In this scenario, security personnel must be available in large numbers to screen passengers at individual gates, and airport operators must provide law enforcement officers at each gate. The disadvantages of this arrangement include high personnel costs and significant delays if the aircraft arrives late because passenger screening cannot begin until the aircraft is available for boarding. This arrangement is mostly used at small airports that handle few flights.

A number of situations during predeparture screening may require the presence or intervention of a law enforcement officer. However, the final decision and the legal responsibility for boarding a passenger rest with the air carrier. If a passenger is arrested for a violation, the air carrier has no further responsibility. If a passenger is cleared for boarding by a law enforcement officer, the air carrier may still decide not to allow the passenger to board.

Role of Aviation Industry Trade Associations

The Air Transport Association (ATA), founded in 1936, is the first and only trade association for the principal U.S. airlines. The ATA represents the industry before Congress, federal agencies, state legislatures, and other governmental bodies and serves as the focal point for industry efforts to standardize practices and enhance the efficiency of the air transport system. Several other aviation trade associations play important roles in maintaining aviation security (Table 3-2).

[1] This number includes passengers, airline crews, airport employees, and people entering to meet or greet arriving or departing passengers.

TABLE 3-2 Aviation Industry Trade Associations

Trade Association	Description
International Air Transport Association (IATA)	IATA is the world trade organization of 250 scheduled airlines. Its members carry more than 95 percent of scheduled international air traffic under the flags of 135 independent nations. IATA produced a security manual to provide airline personnel with reference material, guidelines, and information to carry out their duties.
Airports Council International-North America (ACI-NA)	ACI-NA provides a forum for the exchange of ideas and information to promote cooperation between all elements of the North American civil aviation industry. The primary goal of ACI-NA is the development and improvement of safe, efficient, and economical airport facilities and services and the enhancement of airport capacity. ACI-NA presents members' views and recommendations to governments, industry, and the general public.
Airports Council International (ACI)	ACI, a federation of six regions headquartered in Geneva, Switzerland, consists of some 460 international airports and airport authorities operating 1,250 airports in more than 150 countries and territories. Global issues are addressed through the integrated ACI structure; each region responds to the needs and concerns of airport operators in that region. ACI represents the world's airports in interactions with ICAO, with which it has observer status, and other world bodies.
American Association of Airport Executives (AAAE)	AAAE, the largest professional organization for airport executives in the world, was founded in 1928. The AAAE represents thousands of airport management personnel at public-use airports, which enplane 99 percent of the airline passengers in the United States.
Regional Airline Association (RAA)	RAA represents U.S. regional airlines, as well as suppliers of products and services that support the industry, before the U.S. Congress, Federal Aviation Administration, U.S. Department of Transportation, and other federal and state agencies. Founded in 1975 (as the Commuter Airline Association of America), RAA member airlines transport between 90 and 95 percent of all regional airline passengers.

INTERNATIONAL CIVIL AVIATION ORGANIZATION

In 1944, representatives of 52 nations gathered in Chicago to create the framework for world civil aviation. On December 7, 1944, the Convention on International Civil Aviation (the Chicago Convention) was signed. Its objective was to ensure the safe and orderly growth of international civil aviation and to promote safety. Today, 183 nations have ratified the Chicago Convention and have thus become contracting states.

The International Civil Aviation Organization (ICAO) was established by the Chicago Convention as a vehicle for international cooperation in all aspects of aviation. In 1947, the ICAO became a specialized United Nations agency, headquartered in Montreal, Quebec, with regional offices in Bangkok, Cairo, Dakar, Lima, Mexico City, Nairobi, and Paris. The ICAO consists of an assembly, a council, and a secretariat. The assembly, which is composed of representatives of all contracting states, is the sovereign body of the organization. It meets every three years, reviews the work of the organization, sets policy, and votes on a triennial budget. The council, composed of 33 contracting states elected by the assembly for a three-year term, is ICAO's governing body. The council adopts standards and recommended practices and incorporates them as annexes to the Chicago Convention. The secretariat, headed by a secretary general, consists of professional personnel recruited on a broad geographical basis to conduct the technical work of the organization.

4

Baggage Handling

The objective of this panel is to assess technologies deployed to improve aviation security, both to protect passenger aircraft from explosives and to protect aircraft from damage from an onboard explosion. Explosives can be placed aboard an aircraft via several vectors, including baggage. In this chapter the typical processes for handling carry-on and checked baggage in a secure environment are described, including the use of unit-loading devices (ULDs) as a basis for the discussion of the operational issues for using HULDs (hardened unit-loading devices) as part of an overall aviation security plan (e.g., TAAS).

The six *typical* vectors for introducing explosives are: passengers (on person); passenger carry-on baggage; passenger checked baggage; cargo originating from known, unknown, or consolidated shippers; courier bags; and mail. More subversive vectors include: crew members (e.g., pilots or flight attendants); an intentional or accidental security bypass; food catering service or meal cart; duty-free items; cleaning crew; and service crew (e.g., mechanics, fuelers, baggage handlers). To prevent the introduction of an explosive, all of these vectors must be secure. However, the focus of this panel is on passenger carry-on and checked baggage. The security devices used to prevent the introduction of explosives via the *typical* vectors are shown in Figure 4-1.

MOVEMENT OF BAGGAGE AND CARGO

Approximately 50 percent of all passenger baggage is carried onto airplanes as carry-on baggage; the other 50 percent is checked at the curb, at the ticket counter, or at the gate. The actual distribution of baggage varies by type of aircraft (see Box 4-1). All carry-on baggage is screened at a security checkpoint by an x-ray scanner prior to being brought aboard an aircraft; in some cases, bags are further investigated with a trace explosives-detection device or searched physically (Figure 4-2). Once aboard an aircraft, carry-on baggage is stowed by the passenger in an overhead bin or under a seat.

Checked baggage is sent to a bag room where it is sorted in a variety of ways, depending on the airline and airport.

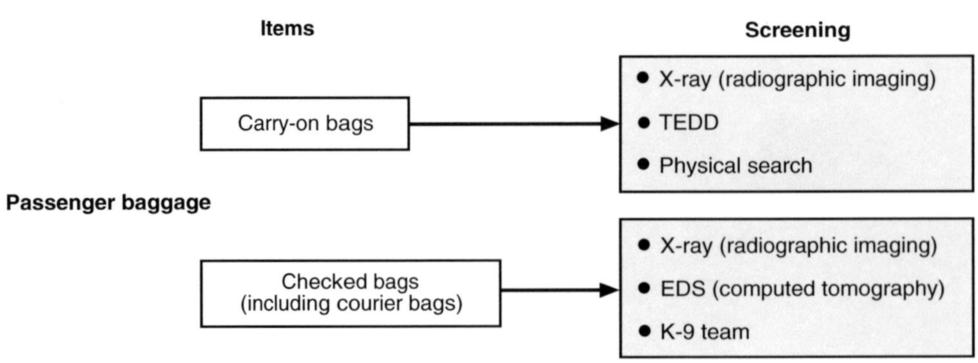

FIGURE 4-1 Vectors for introducing explosives and screening tools.

BOX 4-1
Baggage Distribution

All Aircraft

- 60 percent of baggage travels in narrow-body aircraft (e.g., Boeing 737, MD-80)
- 40 percent of baggage travels in wide-body aircraft (e.g., Boeing 747, 767, 777, MD-11)
- 50 percent of all passenger baggage is checked
- 50 percent of all passenger baggage is carry-on
- 80 percent of all passenger baggage travels as bulk (i.e., loose, noncontainerized)
- 20 percent of passenger baggage travels in containers (i.e., ULDs)

Narrow-Body Aircraft

- 50 percent of passenger baggage travels as bulk on the passenger deck in overhead bins and under seats
- 50 percent of passenger baggage travels as bulk in the cargo hold

Wide-Body Aircraft

- 45 percent of passenger baggage travels as bulk on the passenger deck in overhead bins and under seats
- 55 percent of passenger baggage travels in containers (ULDs) in the cargo hold

The panel observed the sorting and loading of baggage and cargo by more than 10 airlines at Los Angeles International Airport, San Francisco International Airport, and John F. Kennedy International Airport for various types of aircraft and for domestic and international destinations. The purpose of these observations was to assess the synergy of the baggage-handling system with planned screening procedures and the feasibility of using HULDs.

Most of the time, checked baggage is sorted either manually or by automated card readers and routed to the bag "make-up" area for the appropriate flight. In the make-up area, the bags for a particular flight are gathered, sorted by class of service and transshipment, and either loaded into a ULD that is then loaded onto the aircraft (containerized method) or loaded manually onto the aircraft one piece at a time using a baggage cart and conveyer-belt system (bulk method). The bulk method is mainly used for narrow-body aircraft and the containerized method for wide-body aircraft. However, both methods are sometimes used for both types of aircraft. For example, ULDs are used for a few narrow-body aircraft, such as some Airbus A320 and Boeing DC-8 aircraft.

Passenger with a Checked Bag on a Domestic Flight

A passenger for a domestic flight can check bags at the curb, the ticket counter, or the gate. If the bag is checked at the ticket counter,[1] the passenger is asked three questions pertaining to the contents and control of the bag (Figure 4-3). The passenger is also subjected to CAPS (computer-assisted passenger screening). If the passenger is determined by CAPS to be a *selectee,* he or she is also subject to PPBM (positive passenger-bag matching). The bag will then be loaded directly onto the plane if it is a narrow-body plane or placed in a ULD and loaded onto the plane if it is a wide-body plane. If the bag is checked at the gate, PPBM and CAPS are not used. However, bags checked at the gate will have been screened by x-ray radiography and, possibly, trace explosives-detection equipment (Figure 4-4).

Passenger with a Checked Bag on an International Flight

A passenger for an international flight usually checks bags at the ticket counter. The passenger is asked questions pertaining to the contents and control of the bags and is subjected to CAPS and PPBM (Figure 4-5). Checked bags are then subject to examination by an explosives-detection device or a certified EDS or are physically searched.[2] The bags

[1] Passengers who check in at the curb or gate are also asked security questions. If a passenger trying to check bags at the curb is determined to be a selectee, he or she is asked to check the bags at the ticket counter.

[2] Typically this is only applicable to the departure city (the city from which the plane departs for a foreign country) and not for baggage that originates on a domestic flight and is transferred to an international flight.

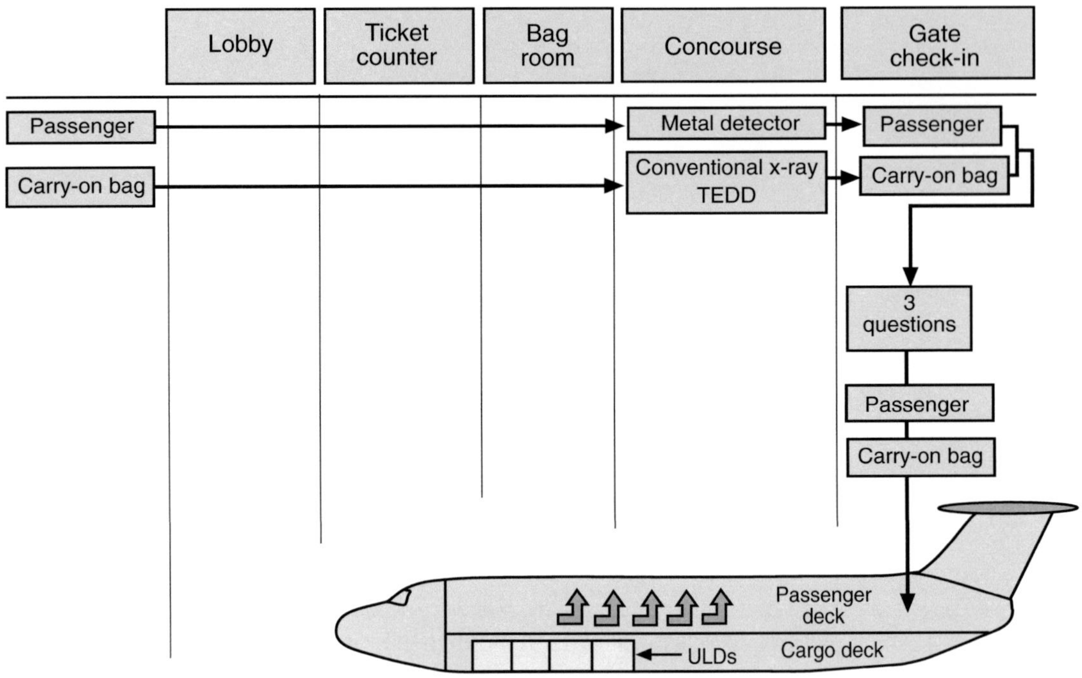

FIGURE 4-2 Baggage flow and screening for a passenger with only carry-on baggage for a domestic flight.

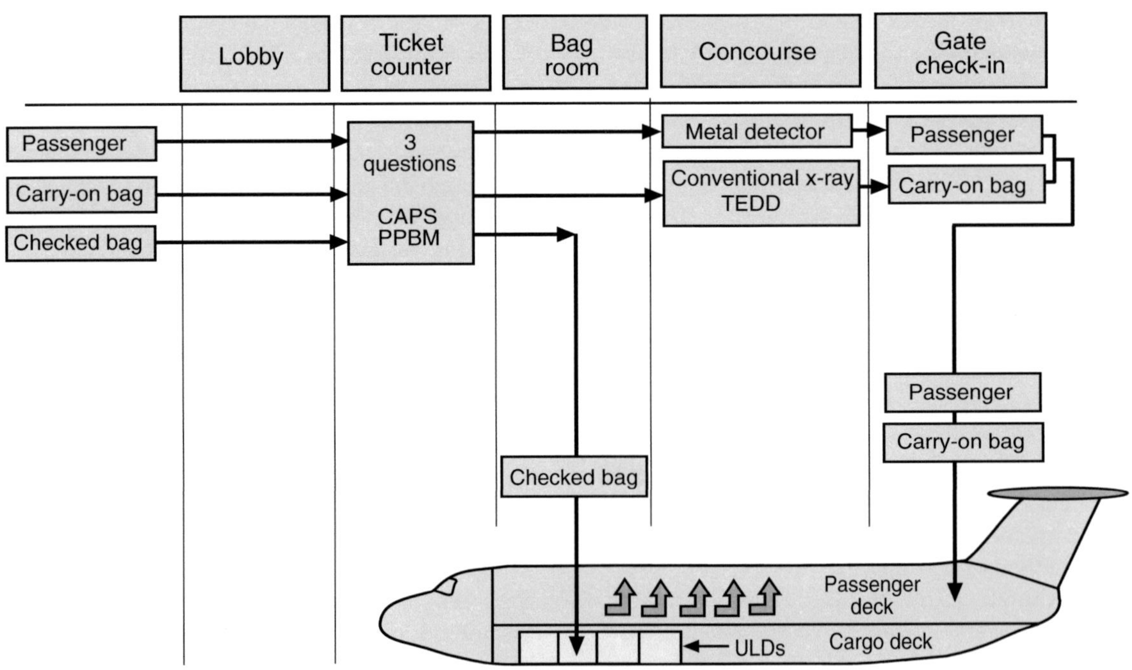

FIGURE 4-3 Baggage flow and security screening for a passenger with a carry-on bag and a checked bag for a ticket counter check-in for a domestic flight.

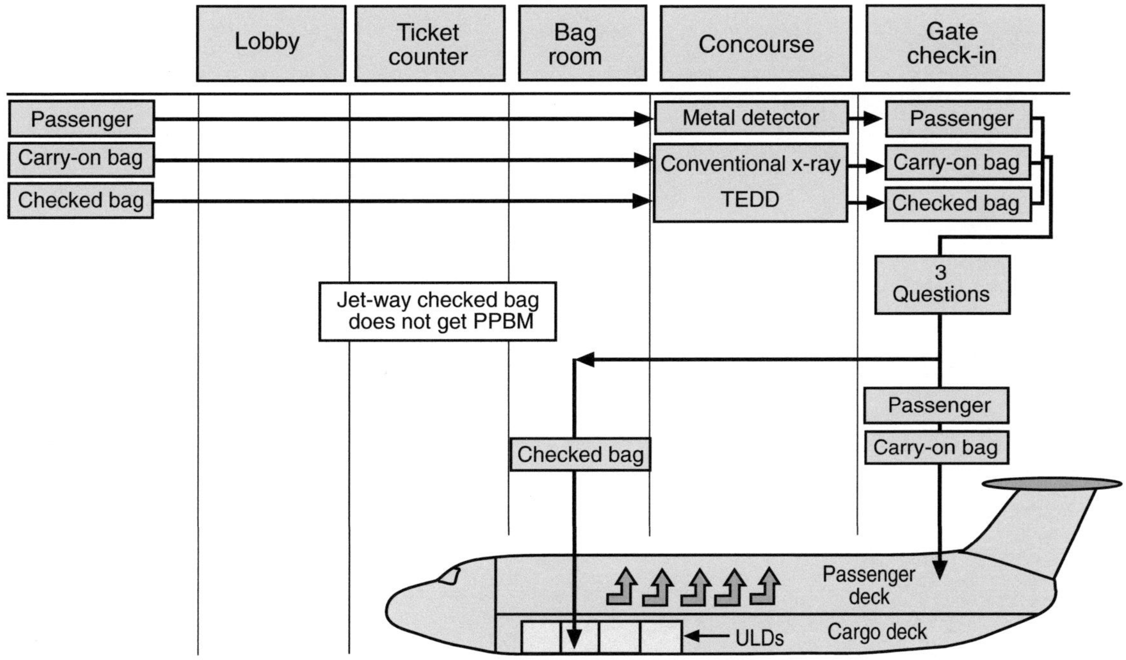

FIGURE 4-4 Baggage flow and security screening for a passenger with a carry-on bag and a checked bag for a gate check-in for a domestic flight.

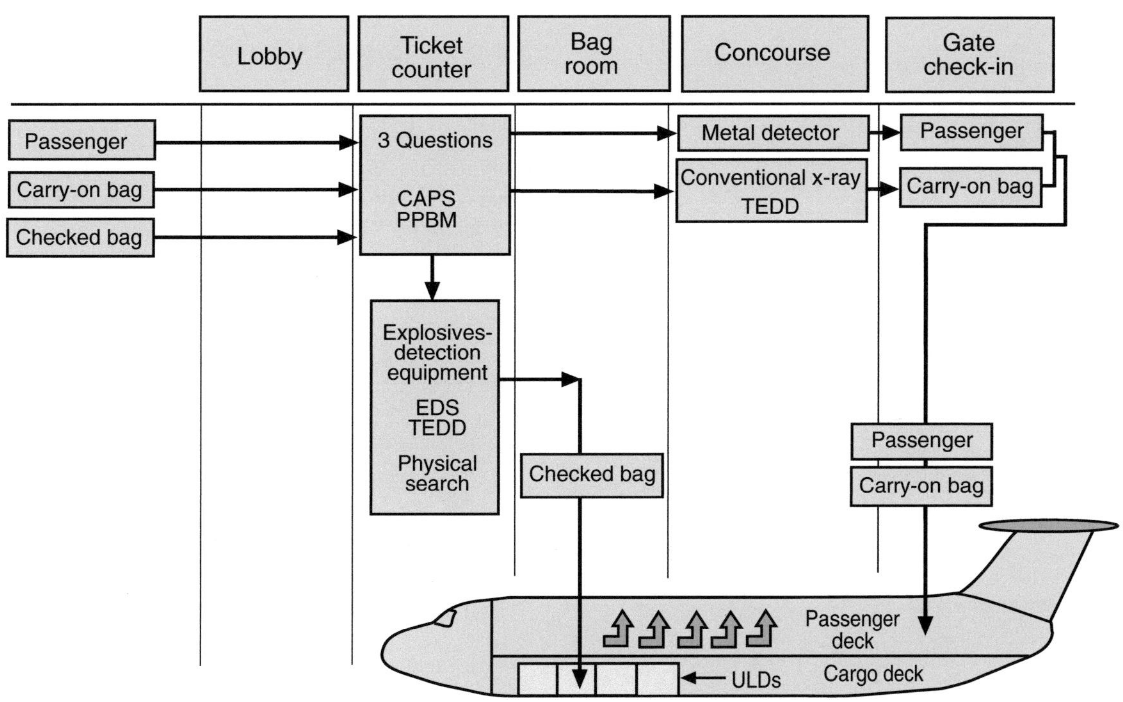

FIGURE 4-5 Baggage flow and security screening for a passenger with a carry-on bag and a checked bag on an international flight.

FIGURE 4-6 Typical LD-3 container.

are then loaded directly onto the plane if it is a narrow-body plane or placed in a ULD and then onto the plane if it is a wide-body plane.

UNIT-LOADING DEVICES

The containerized luggage system is used mainly on wide-body aircraft to facilitate the rapid movement and organization of large quantities of baggage and cargo, minimize on-loading and off-loading times, and facilitate transshipments. The basic components of the containerized luggage system are ULDs, also referred to as containers or cans (see Figure 4-6). ULDs are used to hold, separate, load, and unload passenger baggage and cargo. They come in a variety of sizes and materials, depending on the application (e.g., type of load [baggage or cargo], type of aircraft, deck location, and transshipment needs). ULD sizes are distinguished mainly by the type of aircraft and are designated by the "LD" nomenclature. The construction material of these containers varies from aluminum to polycarbonate to heavy-duty cardboard. In all, there are an estimated 200 or more combinations of ULD sizes, materials, and manufacturers. The containerized luggage system also includes specialized container carts and aircraft loading devices.

Use of Hardened Unit-Loading Devices

During the panel's airport visits, several procedures and practices were observed that could affect the combined use of security screening devices and HULDs within a defined TAAS. The method of loading ULDs varies from airline to airline, and even from airport to airport by the same airline. Some airlines stand the bags up on the floor of the ULD (Figure 4-6); others lay them flat. All airlines observed by the panel tried to maximize the bag count in each container (i.e., load all the way to the top). However, the bag count can vary from 30 to 50 bags per container, depending on the destination of the flight and the average size of the bags (bag size appears to be a function of the local passenger culture and destination). Most airlines separate bags by transfer destination, as well as by class (first class, business class, and coach class). The first-class and business-class ULDs are loaded last and unloaded first. One airline, however, used a separate ULD for the bags of CAPS selectees, which was loaded last.

All ULDs look and function in similar ways. The location of the ULD doors and accessibility for loading are also similar. If the airline route (e.g., Los Angeles to Sydney, Australia) is extremely weight sensitive, only lightweight (e.g., cardboard) ULDs are used. Airlines are very sensitive to the weight issue. For example, contractual requirements with the U.S. Postal Service dictate that any weight overload will usually result in the removal of passengers rather than the removal of mail.

Seventy-five percent of the commercial airline fleet is narrow-body aircraft, which means that 80 percent of baggage travels in bulk cargo holds or in the passenger cabin and only 20 percent in ULD containers. Therefore, if HULDs were used for 100 percent of the wide-body fleet, they would still only account for approximately 20 percent of all passenger luggage. The development and use of HULDs are discussed in Chapter 5.

5

Blast-Resistant Containers

Commercial aviation can be protected from the threat of explosives in two ways, either by preventing explosives from reaching the aircraft (e.g., by using explosives-detection technologies) or by mitigating the effects of an explosive by protecting the aircraft from an onboard explosion (e.g., via aircraft hardening and hardened containers). A combination of these two approaches may provide the best protection of commercial aviation. In this chapter the panel discusses the development of aircraft-hardening concepts in general and blast-resistant (hardened) containers in particular.

ONBOARD EXPLOSIONS

Despite more than 50 bombing incidents in the 25 years prior to the 1988 terrorist bombing of Pan Am Flight 103, the U.S. government and the commercial aircraft manufacturers of the time (i.e., Boeing, Lockheed, McDonnell Douglas, and Airbus) had limited knowledge of the characteristics of internal bomb blasts and the related vulnerabilities of commercial aircraft. All that was known was that blast forces exceeded the design loads of the aircraft, which were never designed to survive a bombing attack. Little, if any, blast data were available, and few analysis techniques were applicable to commercial aircraft. Furthermore, not much was known about protection techniques.

As a result of nearly 10 years of joint FAA, Department of Defense, and industry research, several tools and resources have been identified and, in some cases, developed to characterize internal blasts and to measure the response of aircraft to certain types of bomb blasts. Data are now available defining the effects of "bare-charge" explosives in narrow-body aircraft (via both testing and analysis) and suitcase-contained explosives in both narrow-body aircraft (via testing only) and ULDs in wide-body aircraft (via testing only) (National Institute for Aerospace Studies and Services, 1996; McDonnell Douglas, 1997). The characteristics of bomb blasts and fragmentation from suitcase-contained explosives and suitcase-contained explosives in fully loaded containers have been defined through extensive testing. And the behavior of fuselage skin panels and joints under high-strain rates (i.e., under bomb blast conditions) is now better understood.

As the commercial aviation industry has become much more aware of internal blast-damage mechanisms, it has begun to identify methods for protecting aircraft. This new information has already been a factor in design considerations for reducing the vulnerabilities of aircraft systems and structures to an internal blast in the cargo hold areas of new commercial aircraft. Despite this progress, viable near-term solutions have yet to be demonstrated for protecting the current commercial fleet through retrofitting. Current aircraft designs are, however, already fairly resistant to internal explosions—as evidenced by the 57 percent survival rate of aircraft for all in-flight bombing incidents (35 events) in the past 25 years (Schwartz et al., 1995). Nevertheless, a single bombing event that results in the loss of life or property is not acceptable to the airlines, the economy, or society as a whole.

HARDENED CONTAINERS

Although the direct hardening of aircraft does not appear to be feasible for the current fleet, hardened containers are being investigated by the FAA and other international airworthiness authorities as an alternative near-term solution. Using HULDs (hardened unit-loading devices) to protect aircraft from explosive attacks is not a new concept. Airlines that operate in high-risk areas of the world have been using custom-built containers since well before the Pan Am 103 tragedy. However, because these containers are much too heavy for general use, only one or two are used per aircraft for carrying select items.

As a result of the back-to-back bombings of two wide-body aircraft—Pan Am 103 in 1988 and UTA Flight 772 in

1989—international accident investigation organizations, congressional committees, and other government authorities—began to promote research into methods of hardening aircraft and ULDs against explosives. In the United States, the FAA began funding research on hardened containers, including blast characterization analyses and the development of blast-tolerant design concepts. The basic goal was to develop operationally feasible HULDs that could withstand the blast and fragmentation forces of an explosion.

Initial operational concepts were based on the idea that an aircraft could be protected by using blast-resistant HULDs in place of all standard ULDs. However, for this idea to be feasible, HULDs would require similar cost, weight, and operational capabilities as standard ULDs.

Design Guidelines and Procurement

Following Pan Am 103, the British Air Accident Investigations Branch recommended that "airworthiness authorities and aircraft manufacturers undertake a systematic study with a view to identify measures that might mitigate the effects of explosive devices" (British Air Accident Investigation Branch, 1990). Major initiatives on hardened containers were begun in the early 1990s, when the FAA launched a series of concept studies, including analyses, testing, and development, with approximately 15 different contractors and support organizations (similar efforts were undertaken in Europe). Studies included government-funded research and development by the U.S. Navy, U.S. Air Force, U.S. Army, Boeing, Northrop, Jaycor, Galaxy Scientific Corporation, the Great Lakes Composites Consortium (GLCC), and others. In addition, many companies initiated internally funded independent efforts on the assumption that hardened containers represented an emerging market area. Research and development continued through the mid-1990s and produced data on blast characteristics, material properties, the feasibility and effectiveness of HULD concepts, and modeling and simulation requirements.

After about five years, however, because of slow progress on an effective, feasible design concept—and because of a less than enthusiastic response from the airlines—no major breakthroughs had been made. Although many concepts were investigated, none was identified as a "silver bullet" that would immediately replace current ULDs. Research did not lead to a hardened container that looked, behaved, and cost the same as a standard ULD, and airlines were not willing to support concepts that would add weight, cost, or operational constraints to their infrastructures.

Although the FAA has considered many approaches to deploying HULDs, in the past few years the focus has been on HULDs as an integral component of the entire security system that would be deployed at the level of one or two per aircraft in conjunction with passenger screening and CAPS (this approach is consistent with the TAAS approach recommended in this report). Only a small percentage of passenger bags carried in the HULD would be based on predetermined selection criteria. The FAA hoped this approach would be acceptable to the airline industry because it would reduce the requirements for the deployment of HULDs.

In 1996, the FAA announced the first of two procurement solicitations for hardened containers that could contain an explosion from a defined explosive threat size and could be deployed into normal airline operations. Six proposals were submitted by various suppliers. However, after determining that none of the proposed concepts would meet the design blast criteria, the FAA decided not to award any contracts (Hacker, 1998a). After redefining the requirements, the FAA released a second solicitation in January 1997 (FAA, 1997d). The requirements for the second solicitation are given in Box 5-1. The FAA received eight proposals in response to the second solicitation, and two proposals, by Jaycor and Galaxy, were selected for development and testing. Both were required to demonstrate compliance with the standards listed in Box 5-2.

Testing

Under the terms of the FAA HULD contracts, Jaycor and Galaxy were contracted to provide prototype HULDs that could withstand the blast effects from a bomb of a certain size and configuration (the blast-resistance criteria for HULDs are classified [ISO 6517]). Each company was required to demonstrate the survivability of its HULD design either through a blast test or a simulation analysis. Both companies decided to meet this requirement through blast tests.

Prior to 1997, the FAA and industry had conducted numerous blast tests on earlier designs and had developed test standards (ISO 6517) for the two candidate HULD designs. The HULD test program, which was initiated in late 1997 in Tucson, Arizona, was performed in a salvaged, unpressurized, lower cargo-hold section from a Boeing 747 aircraft that included the basic aircraft structure (e.g., airframe, stringers, skin panels, shear clips, wall liners, cargo floor beams, and passenger floor beams), as shown in Figure 5-1. Although most of the major aircraft systems had been removed, much of the wiring, cables, and air ducts were intact. Each HULD was placed separately into the cargo-hold area at a standard container lock-down location. No other containers were included in the tests. Test objectives included verification (1) that the HULD could withstand the explosive blast with no rupture or fragmentation; (2) that no fragments penetrated the HULD or the aircraft structure; (3) that the HULD had not moved or jumped causing structural damage to the aircraft (i.e., airframe, cargo floor, or passenger floor); and (4) that no fire had erupted after the blast.

BOX 5-1
FAA 1997 Solicitation for Hardened Containers (DTFA03-97-R-00008)

- Design and develop a blast-mitigating HULD that would contain the classified threat level.
- The HULD must meet normal aircraft ULD interface standards.
- Prove the HULD will survive the threat, through either test or analysis.
- Supply the FAA with up to 60 of the HULD containers for test deployment on a selected (volunteer) air carrier.
- Provide design details, development concepts, prototype HULDs, testing support, operations/maintenance manuals, and deployment support and repair.

The blast tests took place at a private facility adjacent to Davis-Monthan Air Force Base in Tucson. Four different HULD concepts were tested: the Jaycor and Galaxy HULDs, a third HULD concept developed by Contemporary Products (funded independently), and an FAA-sponsored "foam-offset" protection method using a standard aluminum LD-3 container. Table 5-1 provides basic descriptions of the four containers.

The first container tested was the Galaxy HULD concept in October 1997. The container failed the test when the floor base separated from the wall, mainly along a manufacturing joint. Galaxy reviewed the failure mode, modified the design, and performed a second test with the FAA in March 1998. This time the HULD survived the blast test (see Figures 5-2 and 5-3). The blast caused some external deformation of the container and some minor deformation to one of the 747 passenger floor beams but no serious structural damage to the aircraft. The temperature inside the HULD peaked at 400°F, after which fire-safety personnel applied water to the HULD. The HULD contained the blast, fragments, and also any potential fire. This Galaxy container meets all of the FAA design and blast criteria and has been certified. The FAA Transport Airplane Directorate Certification Office documented the certification on July 29, 1998 (FAA, 1998). A second Galaxy Scientific design passed the FAA blast test required for certification in January 1999.

The first prototype of the Jaycor HULD was tested in March 1998. In this test, the HULD failed because of separation of the composite ceiling material and the failure of the container door. Jaycor reviewed the failure mode and modified the prototype for a second test that was conducted in January 1999. However, the HULD again failed the blast test because of a failure around the door area.

Blast tests were also conducted on a HULD developed by Contemporary Products. This container, developed with funding from the state of Wisconsin, was made of an E-glass-based composite and weighed 1,100 pounds (four times more than a standard ULD). When the unit was tested, the HULD split at a manufacturing line, and the door separated in an explosive manner. Although the unit itself failed the test, there was minimal damage to the airframe structure. In flight, however, damage could have been much more extensive because of the presence of other containers and pressurization.

The foam-offset concept consisted of a standard aluminum LD-3 with a 12-inch-thick rigid-foam block on the sloped panel of the ULD. This configuration prevents the placement of luggage adjacent to the sloped wall of the LD-3, thus ensuring standoff distance between an explosive device and the aircraft skin panels to reduce shockholing and, potentially, blast forces. The foam offset was originally tested by the British in the Bruntingthorpe test and showed very promising results (Morrocc, 1997). The idea was especially attractive because it was low cost, added little weight, and required little retrofitting.

For the FAA test, the explosives-containing suitcase was placed near the bottom of the container close to the sloping panel. The explosive blast and attendant fragmentation ripped through the foam offset, the LD-3 aluminum wall, and the skin of the aircraft. It also caused some damage to

BOX 5-2
Standard HULD Requirements

- Conform to NAS-3610-2K2C certification (basic design and load requirements).
- Conform to TSO 90-C (flame test, heat vs. time).
- Conform to draft ISO Standard 6517 (blast and fire requirements):
  - — test for blast (shockholing and fragmentation) and fire containment
  - — decompression requirements
  - — no collateral damage to aircraft
- Receive FAA letter for HULD engineering design approval.
- Maintain blast mitigation ability after airline "in-service" trial.

FIGURE 5-1 Cargo hold for blast testing HULDs.

the floor beams. Thus, despite the positive results from Bruntingthorpe, the test under the FAA blast criteria was an obvious failure. Table 5-2 summarizes the test results.

Deployment

Even though a certified HULD is now available, the FAA has not developed a deployment plan for airline service. The idea of replacing all existing containers with HULDs has been abandoned because of its impracticality in terms of weight, cost, and operational factors. The alternative deployment scenarios range from 100 percent utilization on international flights to single-container usage on all flights or on selected flights. Table 5-3 shows the estimated costs of the single HULD per aircraft scenario. The cost to cover the fleet for a single HULD per plane is estimated to be $125 million in acquisition costs plus an unknown recurring cost, which could be substantial. If all standard ULDs were converted to HULDs, the acquisition costs alone could reach $2.5 billion.

If a single HULD is used per flight or for selected flights, the contents of the HULD would have to be carefully selected. Contents could include the following items:

- selectees' bags from CAPS
- selected bags from cleared EDS alarms
- unaccompanied luggage
- mail and other cargo

The FAA reports that an initial deployment plan will be developed after the initial purchase of several Galaxy HULD units and a test deployment with an airline (Hacker, 1998b). The test deployment will enable the FAA to assess operational implications, such as maintenance requirements, repairability, and baggage flow. The FAA will also be able to determine if a HULD maintains its blast strength after

TABLE 5-1 Characteristics of HULDs Tested

	Galaxy	Jaycor	Contemporary	Foam Offset
Skin material	GLARE (aluminum/fiberglass composite)	Composite	E-glass composite	Aluminum
Joints	Aluminum frame/rivets	Integral/molded	Unknown	Aluminum frame/rivets
Weight	330 lb.	450+ lb	1,100 lb.	250 lb.
Cost (1,000+ units)	$20,000	Unknown	Unknown	$2,000

FIGURE 5-2 Galaxy HULD in test position prior to blast test.

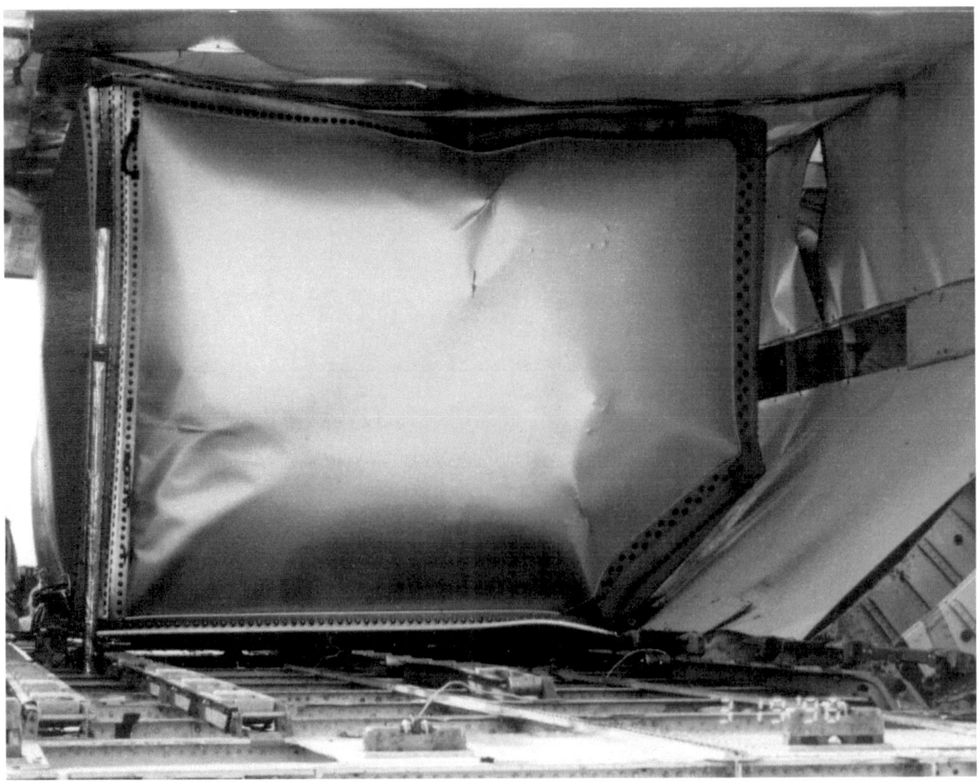

FIGURE 5-3 Galaxy HULD after blast test.

TABLE 5-2 Summary of HULD Test Results

	Galaxy			Jaycor			
	I	II	III	I	II	Contemporary Products	Foam Offset
Test date	Oct 97	Mar 98	Jan 1999	Mar 98	Jan 99	Mar 98	Mar 98
Results	Failed	Passed	Passed	Failed	Failed	Failed	Failed
Failure Mode	Floor separation	—	—	Door failure	Door failure	Joint separation	Blast hole

being subjected to the typical wear and tear of airline operations. Originally, the FAA planned to purchase and deploy as many as 60 containers for this assessment. However, because of program changes and budget constraints, the number has been reduced to 10, which are currently being tested by two air carriers.

Operational Issues

The concerns listed below have been raised by the airline industry through the International Air Transport Association (IATA) and the ATA about the cost and operational aspects of HULDs (Rork, 1994; IATA, 1995):

- HULDs weigh more than ULDs (weight affects aircraft range and payload capacity).
- HULD procurement costs are higher than ULD costs.
- HULDs may require extra maintenance (e.g., nontraditional materials, training, and testing).
- Certain HULDs could reduce cargo volume (which could cause a loss of revenue).
- Airline costs could be higher because of extra weight and maintenance.
- HULDs may not be available for every flight if maintenance problems arise.

In addition to these general concerns, the airlines have also questioned the utility of the recently certified HULD design, which some have called "inadequate" because of the design and location of the door, which uses the entire inboard (aircraft centerline) side of the container and swings open as a single hinged unit that sweeps a large arc. This rear-door design is not compatible with the design of airline bag rooms, bag carts, and loading processes. Despite these problems, some carriers say they may be amenable to using HULDs if they are limited to one per aircraft.

The FAA has procured 10 HULDs in the current configuration (rear door) from Galaxy for airline deployment. Three airlines volunteered to use these HULDs for a one-year trial period. In addition, the FAA procured a new HULD prototype (from Galaxy) with a side door (the traditional location) that passed the blast certification test in January 1999.

TABLE 5-3 Panel's Estimated Costs for the Procurement and Operation of 12,500 HULDs

	Annual Recurrent Costs	Nonrecurrent Costs	Notes
Acquisition costs	0	$125 million	One HULD per airplane (roughly one of every 20 ULDs would be replaced by a HULD) for a total of 12,500 HULDs deployed. Assume each HULD costs $10,000.
Operation, maintenance, and support	TBD	0	
Increased aircraft fuel consumption due to increased weight	$6 million	0	HULDs weigh more than most ULDs.
Lost revenue due to increase in operating empty weight	$5 million	0	20 percent of long-haul flights (annually) would lose revenue.
Lost volume	0	0	HULDs are not expected to have less storage volume.

CONCLUSIONS AND RECOMMENDATIONS

Overall, the FAA has conducted a reasonable effort in sponsoring and driving the development of HULDs. As a result of FAA/industry efforts, a certified airworthy HULD design is available for deployment. Nevertheless, the panel believes that significant improvements could be made in HULD designs, testing, and operations.

The panel's greatest concern is that research and development on HULDs have not been conducted on an SOS (system-of-systems) approach. The HULD has largely been developed and designed as a single stand-alone entity, and limited research has been done on its role and integration into a TAAS (total architecture for aviation security). Coordination with the airlines, airports, and aircraft manufacturers has been mainly in the areas of specific design and utility requirements. The FAA and airlines have not focused on the interactions, boundary conditions, and trade-offs of using HULDs along with other security measures, such as passenger profiling and baggage screening. The panel believes that alternative, more practical, HULD designs could still be developed.

Recommendation. The FAA should not deploy hardened unit load devices unless they are determined to be a necessary security feature of the total architecture for aviation security (TAAS), as determined by the FAA and other stakeholders, on the basis of cost, operational, and deployment studies.

Recommendation. The FAA should go forward with the planned operational testing of hardened unit load devices.

HULD Design

The FAA's approach to HULD design has been focused primarily on three areas: airworthiness, ground handling, and blast resistance. Little attention has been given to operational and support (O&S) issues or systems-integration issues. O&S considerations, which, include inspection, certification, and repair, should be addressed concurrently with HULD development. These issues are very important, even in the early development phase. A HULD that cannot be supported or repaired would be essentially useless to the industry. Therefore HULDs should be "designed for supportability."

The research and development of hardened containers for narrow-body aircraft are lagging far behind the work on containers for wide-body aircraft. This panel believes that a HULD for narrow-body aircraft must be available before the FAA recommends, mandates, or regulates the use of hardened containers for airline operations. Both narrow-body and wide-body aircraft are used on many international routes, especially to Central and South America, and even on several routes between the United States and Europe. Flying narrow-body aircraft without hardened containers and wide-body aircraft with hardened containers on the same routes would be analogous to flying certain types of aircraft over water without "life vests." The risk is still the same, but the level of protection is lower.

Recommendation. The FAA should continue to support research and development on hardened unit-loading devices. In addition to performance, operational considerations (e.g., operability, supportability, inspection, and repair) should also be addressed.

Recommendation. Before the FAA recommends, mandates, or regulates the use of hardened containers for airline operations, the issue of containing explosions aboard narrow-body aircraft must be resolved. The FAA should pursue the development of hardened unit-loading devices for narrow-body aircraft.

HULD Testing

The FAA has successfully tested two HULD prototypes produced by Galaxy Scientific, in the cargo hold of a salvaged aircraft. One had a rear-door configuration, the other a side-door configuration. Both HULDs survived the blast, and no fire was observed. In the panel's opinion, certifying a HULD on the basis of one blast test is not credible for the following reasons:

- The FAA requires "self-extinguishing" of any fire initiated in a HULD, but fire safety (in the panel's opinion) cannot be verified by a single test.
- A single full-scale blast test (in addition to component and material tests) does not rigorously test the HULD. Although there is a fairly good understanding of test requirements and procedures, single-test certification is questionable because of the variability in test parameters, such as bomb placement in the HULD, explosive charge performance, differences in luggage, and differences in luggage contents.

The FAA could remedy this situation in two ways. First, a more rigorous test plan could involve a series of tests in a pressurized aircraft (or representative test facility) over a range of conditions (e.g., explosive configurations, baggage-load scenarios). Second, Monte Carlo analyses (i.e., probability distributions to estimate the outcome of a statistical event) could be used to test a range of conditions.

Recommendation. The FAA should implement a more rigorous test plan for certifying hardened unit-loading devices. The plan should include a series of blast tests, as well as modeling and simulation (e.g., Monte Carlo analyses). To carry out this test plan, the FAA will probably have to improve its modeling and simulation capability and construct more robust testing facilities.

HULD Deployment and Operation

Although the FAA has recently certified two HULD prototypes and initiated a limited deployment, no comprehensive plan for integrating HULDs into the container logistics system of an airline has been developed. Issues that remain to be addressed include type of aircraft, number of HULDs per aircraft, and the type of baggage or cargo put into HULDs.

A deployment strategy to use HULDs to transport the bags of CAPS selectees and bags cleared by an operator after an explosives-detection equipment alarm raises legal issues. On the one hand, airlines have expressed concerns about their liability if a container (i.e., a HULD) known to contain bags considered more of a threat than other bags is put aboard an aircraft. On the other hand, if there were an in-flight explosion resulting in hull loss, the airlines might be held liable for not using available HULD technology.

The deployment strategy described above also raises some logistical issues. Introducing hardened containers into the airline system, especially in small numbers, would reduce aircraft loading flexibility. Airlines currently separate baggage by class of service and destination to expedite handling. This process begins in the baggage make-up area, where specific containers are allocated to different destinations and then separated by class of service. Selectee bags that varied by destination and/or class of service would require either multiple hardened containers or changing the baggage-handling and sorting process.

The hardened containers tested to date exceed the acceptable weight criteria specified by the airlines through ATA (Rork, 1994). Container weight has a significant impact both on ground handling systems and aircraft loading systems. In addition, for structural survivability, a "best location" is identified for HULD(s) aboard each aircraft type, which may conflict with optimum weight and balance requirements. Also, to accommodate late-check-in "selectees," the HULD(s) would have to be one of the last containers loaded onto the aircraft and might require that containers already onboard be off-loaded to ensure proper loading arrangement by destination, weight, and cargo classification. Finally, a means of tracking would have to be developed to ensure that HULDs were available for each flight.

Container weight also has a significant impact on fuel and revenue displacement. IATA's preliminary cost estimates for the procurement and support of HULDs included a fleet-wide annual recurrent cost of more than \$5.0 billion[1] and a nonrecurrent cost of \$600 million. However, in the panel's opinion, some of the assumptions underlying these calculations are not accurate, and the costs are overestimated. The panel's estimate for a single HULD per wide-body aircraft is substantially lower. The recurring O&S costs (which could be very substantial) cannot be estimated yet, but other recurrent costs are estimated to be about \$11 million for an inventory of one HULD per aircraft. The panel estimates that the nonrecurrent cost would be \$125 million.

Recommendation. The FAA should work closely with airlines on the development and deployment of hardened unit-loading devices.

Recommendation. In consultation with the airlines, the FAA should develop cost estimates for recurring and nonrecurring costs to the airlines for the deployment and operation of hardened unit-loading devices (HULDs) based on a single HULD per flight for both wide-body and narrow-body aircraft and for 100 percent deployment.

---

[1] The \$5 billion per year recurring cost projected by IATA reflects lost commerce because of the weight of HULDs. For example, if HULDs increase the weight of an airplane, an equal amount of cargo cannot be shipped on the plane. IATA categorized this loss of revenue as a recurring cost.

6

Bulk Explosives Detection

In this chapter the FAA's progress in deploying bulk explosives-detection equipment and the operational performance of this equipment are evaluated. Bulk explosives-detection equipment includes any device or system that remotely senses a physical or chemical property (or combination thereof) of an object in an attempt to detect (semiquantitatively) the presence of an explosive concealed in a container (e.g., passenger baggage). Arguably, bulk explosives-detection equipment is the foundation of the TAAS in the current environment. The critical performance metrics for explosives-detection equipment include probability of detection (P_d), the probability of false alarm (P_{fa}), and throughput rate. At the same time, this equipment must function without unreasonably interrupting passenger flow.

An FAA-certified EDS (explosives-detection system) is a self-contained unit (composed of one or more integrated devices) that has passed the FAA's certification test. To obtain certification, equipment must meet the following standards:

- The detection rates against various types of explosives contained in baggage must have an overall P_d of no less than *X*[1]
- The P_{fa} (as determined for airline-type baggage) must not exceed *Y*.
- The baggage throughput rate of the equipment must meet or exceed 450 bags per hour.

A number of nuclear and x-ray-based techniques have been investigated, but only three systems (all based on x-ray computed tomography [CT]) have been certified.

[1] The actual values of *X* and *Y* are classified and can be found in classified FAA reports (FAA, 1992).

APPLICATION OF BULK EXPLOSIVES-DETECTION EQUIPMENT TO POSSIBLE THREAT VECTORS

Passengers

Current passenger-screening requirements were developed in 1972 in response to an increase in hijackings (NRC, 1996b). Current passenger-screening procedures involve metal-detector portals that can detect only metallic weapons. Emerging imaging technologies, however, can detect the presence of both metallic and nonmetallic weapons, as well as explosives concealed under multiple layers of clothing (NRC, 1996b). These technologies, which include passive and active millimeter-wave imaging and active x-ray imaging, all require human operators to view and interpret the images. Because the images are somewhat revealing of the human anatomy, passengers are likely to object to the images being displayed to an operator. Furthermore, active imaging techniques require radiation, which raises some health concerns. Because of these concerns about privacy and health, imaging technologies will probably not be deployed for passenger screening at the current threat level (NRC, 1996b).

Carry-on Baggage

At present, all carry-on baggage is screened by conventional x-ray radiography. Manufacturers of explosives-detection equipment are working on new technologies that are just being evaluated by the FAA. Operator-assisted x-ray, for example, highlights areas in a radiographic image that could be a threat object (e.g., a weapon or bomb); the highlighted image is then evaluated by the operator (Polillo, 1998). Because this technology is in its infancy and has not been widely deployed, not enough data were available for the panel to evaluate its performance. Nuclear quadrupole resonance is another possible technology for this application.

However, to date, the range of explosives this technology can detect is limited.

Checked Baggage

Because checked baggage is the threat vector that has received the most attention from the FAA, bulk explosives-detection equipment has become the most critical component of the TAAS by default. To date, three explosives-detection systems have been certified by the FAA, all x-ray CT-based systems manufactured by InVision Technologies, Inc. (CTX-5000, CTX-5000 SP, and CTX5500 DS). Other vendors have developed bulk explosives-detection equipment, but none of them has passed certification testing. In addition to the FAA-certified InVision CTX series EDSs, some noncertified explosives-detection devices have been selected for deployment, including Vivid, EG&G Z-Scan, and Heimann devices. The discussion in this chapter focuses predominantly on findings pertaining to the InVision CTX series EDSs and the available data on the performance of deployed units.

DEPLOYED BULK EXPLOSIVES-DETECTION EQUIPMENT[2]

Based on the recommendations of the White House Commission on Aviation Safety and Security (1996, 1997), Congress mandated the deployment of bulk explosives-detection equipment in U.S. airports. In late 1996, the FAA associate administrator for civil aviation security instructed the SEIPT to assess the availability of explosives-detection equipment for deployment in commercial airports, develop a deployment strategy and plan, and execute the plan. The plan called for the deployment of the FAA-certified InVision CTX-5000[3] series systems, as well as deployment of so-called advanced technology (AT) hardware, namely eight Vivid devices, 10 EG&G Z-SCAN devices, two InVision/Quantum Magnetics Q-Scan devices, and two Heimann devices by December 1997. The first InVision CTX-5000 was deployed in January 1997, and two to 10 deployments were planned for every month during 1997. The FAA/SEIPT's progress as of January 1, 1999, is summarized in Table 6-1; the locations of the equipment in airports are summarized in Table 6-2.

The installations are classified into two types, stand-alone and integrated. Stand-alone installations are divided into four types based on their location: ticket counter, lobby, baggage area, and security area. Integrated installations are divided into two types, output and fully integrated installations. Output installations are configured to run directly into the baggage-handling system and can be located in the lobby, at the ticket counter, or curbside. Fully integrated EDSs are completely integrated (i.e., both input and output) into the baggage-handling system.

Installation costs are dependent on the type of installation and whether the installations require major modifications. For example, the lobby installation of a stand-alone CTX-5000 SP that does not require major airport modifications costs on the order of $10,000 to $30,000. However, a stand-alone lobby installation that requires major modifications, such as reinforcing the floor, cement work, or moving staircases or other structures, can raise the cost to $100,000 or $200,000. For more difficult integrated installations at existing airport terminals, the costs can be even higher. In the United Kingdom, fully integrated CTX-5000 installations account for 25 to 50 percent of the cost of the whole baggage-handling infrastructure. The British Airport Authority has determined that, in general, every $2 million spent installing CTX-5000 EDSs will require approximately $6 million to complete the modifications to the baggage-handling infrastructure. Note that this installation cost is not just relevant to CTX-5000 EDSs. The cost of installation (due to modifications to the baggage-handling infrastructure) of explosives-detection equipment of similar size and weight (e.g., EG&G, Vivid, and Heimann) would probably be comparable.

Most of the bulk explosives-detection equipment that has been deployed, or is being considered for deployment, is x-ray based—the exception being the InVision/Quantum Magnetics Q-Scan, which is based on a nuclear quadrupole resonance measurement technique. The x-ray-based technologies that have been deployed—or are scheduled to be deployed—include transmission x-ray, dual energy x-ray, and CT. Of these, only the CT-based technology has passed the FAA bulk explosives-detection certification test. The others were selected because they have large enough apertures to handle oversized bags, and operational data would be useful. The performance baseline for the noncertified explosives-detection devices has been determined at the FAA Technical Center. The sole electromagnetic instrument scheduled for deployment, developed by Quantum Magnetics, Inc. (recently Quantum Magnetics was bought out by InVision, Inc.), is a nonimaging technique based on nuclear quadrupole resonance (NQR). The conditions specified by the FAA for use of the deployed equipment are shown in Box 6-1.

[2] Note that bulk explosives-detection equipment predominantly addresses the threat of an explosive device being brought aboard an airplane via checked baggage. The current deployment of this equipment does not address other threat vectors, such as a passenger carrying explosives on his or her person, carry-on baggage, cargo, mail, or catered food.

[3] This number includes one CTX-5000 already deployed at San Francisco International Airport and two CTX-5000s deployed at Atlanta's Hartsfield International Airport for previous operational testing.

TABLE 6-1 Planned and Actual Deployments of Bulk Explosives-Detection Equipment

Manufacturer/Model	Initial Planned Deployments	Revised SEIPT Plan (to be deployed by 2/28/99)	Actual Deployments (1/1/99)
InVision CTX 5000 SP and 5500 DS	54	74	71
Vivid VIS-1	10	8	2
EG&G Z-Scan 7	10	10	3
Heimann HI Scan	0	2	1
InVision/Quantum Q-Scan	5	2	0

TEST DATA

To date, most of the performance data on deployed explosives-detection equipment have been generated from tests conducted at the FAA Technical Center. However, operational test data on P_{fa} (false-alarm rates) are also reviewed in this section.

Test Data from the FAA Technical Center

Most performance data for x-ray CT-based EDSs are from certification testing and operational testing of the InVision CTX-5000 SP. Data on the performance of the other FAA-certified EDSs and from the recent certification testing of InVision CTX-5500 DS and L3 Communications 3DX-6000 were not available at the time this report was written. The InVision CTX-5500 DS has been certified for two different inspection modes: SURE98 mode and CERT98 mode. In SURE98 mode, it has a lower P_{fa} but also a lower throughput rate than the CTX-5000 SP. In CERT98 mode, it has a similar P_{fa} to the CTX-5000 SP but a much higher throughput rate. The panel's analysis of performance data focuses on the certified CTX-5000 SP and CTX-5500 DS, although some data on other deployed explosives-detection equipment are also presented. Table 6-1 shows the performance factors for deployed explosives-detection equipment, including the P_d, P_{fa}, and the bag throughput rate. Because the actual P_d and P_{fa} numbers are classified,[4] P_d is given as a percentage of the overall P_d required for certification (X), and P_{fa} is given as a percentage of the P_{fa} required for certification (Y). The InVision/Quantum Magnetics Q-Scan has not been tested at the FAA Technical Center.

Operational Test Data

In 1995, the FAA initiated the Airport Operational Demonstration Project to determine the operational performance of the InVision CTX-5000 SP in the field as compared to its performance in certification testing (FAA, 1995). Three sites were selected for the project: San Francisco International Airport (United Airlines); Atlanta's Hartsfield International Airport (Delta Airlines); and Manila International Airport (Northwest Airlines). This operational demonstration project was not initially related to the congressionally mandated deployment of explosives-detection equipment. Recently, however, the FAA decided to include the CTX-5000s installed during the operational demonstration project in the overall deployment.

Two InVision CTX-5000s were installed in Atlanta and one each in San Francisco and Manila. The demonstration project included four open tests and one blind (so-called "red team") test using improvised explosives devices (IEDs) to determine P_d. The P_{fa} was measured routinely throughout the project on real passenger bags. Only the data from the San Francisco and Atlanta deployments have been documented in final reports (FAA, 1997a, 1997b, 1997c). Some of the performance data from San Francisco International Airport are given in Table 6-3.

The automated explosives-detection capability of the CTX-5000 SP in the field (% $X = 102$) was about the same as the capability measured during laboratory testing (% $X = 106$) at the FAA Technical Center. However, operator intervention to resolve alarms measurably reduced the overall P_d. This tendency was also observed during blind testing at San Francisco and Atlanta. During the operational demonstration project at San Francisco, the automated P_{fa} was 113 to 150 percent higher than the certification standard. For the present study, supplementary data were provided to the panel by SEIPT with P_{fa} from January 5, 1998, to April 20, 1998. These data show that P_{fa} varies between 125 and 250 percent higher than the maximum rate allowed during certification. Operational data reviewed by the inspector general of the U.S. Department of Transportation suggested that the P_{fa} was as high as 169 percent higher than the certification standard (DOT, 1998).

Data from the first of the four open tests show an average of 50 seconds for alarm-resolution time using the CTX-5000 SP. Although the resolution time was lower during subsequent tests, the combination of a high P_{fa} and a long alarm resolution time can have a significant impact on the throughput rate, and, in fact, was determined to be the limiting factor for throughput rate.

[4] The actual values required for certification are recorded in classified FAA documents (FAA, 1992).

TABLE 6-2 Location of Deployed Bulk Explosives-Detection Equipment (April 1999)

Location	InVision CTX	Vivid VIS-1	EG&G Z-scan 7	Heimann HI-Scan	InVision/QM Q-Scan	Total
Lobby stand-alone	40	1				41
Lobby/curbside integrated[a]	19	1		2	1	23
Bag room, stand-alone	3					3
Bag room, integrateda	12		4			16
FAA Technical Center	1					1
To be determined		6	6		1	13
Totals	75	8	10	2	2	97

[a] Screening device is either partially or fully integrated into airline baggage-handling system.

CONCLUSIONS AND RECOMMENDATIONS

Deployment

The deployment of bulk explosives-detection equipment has not progressed as quickly as planned. Initially, 54 certified bulk explosives-detection systems were scheduled to be deployed by December 1997. The deployment plan was then modified, and the EDSs, as well as 22 noncertified bulk explosives-detection devices, were to be deployed by March 1999. In the interim, the FAA developed a program to purchase the equipment along with developed and implemented factory and site-acceptance protocols and testing procedures. As of January 1999, more than 70 certified x-ray CT-based EDSs had been deployed and seven other bulk explosives-detection devices. Therefore, the deployment of more explosives-detection equipment based on other technologies would yield useful operational data.

Recommendation. The FAA should first deploy and obtain operational data on advanced technology (AT), including EG&G Z-Scan, Vivid devices, Heimann devices, and InVision/Quantum Mechanics QR devices. The FAA should then deploy and obtain operational data on any warehoused InVision CTX-5000 SPs or CTX-5500 DSs.

The location of a CTX-5000 in an airport is largely dictated by physical constraints, which in turn can affect its utility. In newly designed airports or airport terminals, the placement of bulk explosives-detection equipment can be incorporated into the design of the terminal (e.g., Terminal One at John F. Kennedy International Airport). After site visits to three airports, the panel concluded that the data are not sufficient to assess the operational installation configuration of bulk explosives-detection equipment in airports.

Recommendation. The FAA should encourage airlines and airports to implement different explosives-detection equipment installation configurations so that their effectiveness can be assessed.

Recommendation. Because foreign airports often use different installation configurations for explosives-detection equipment than U.S. airports, the FAA should collaborate with foreign governments to collect data on the effectiveness of various configurations to assist in establishing the best practices.

Data Collection

The FAA has not developed a plan for collecting data on the P_{fa} and operator alarm resolutions (e.g., actions taken, time to resolve, etc). During three separate site visits, the panel found no evidence of measures being used to assess the performance of deployed equipment or of an FAA-specified data-collection protocol. InVision Technologies has taken the initiative of collecting data on its deployed systems. The panel concluded, however, that the available data were insufficient to evaluate the operational effectiveness of deployed equipment. Furthermore, the panel believes that little data on the operational effectiveness of the deployed equipment will be forthcoming unless the FAA develops a plan with the airlines and explosives-detection equipment manufacturers to obtain such data.

Recommendation. In cooperation with the airlines and explosives-detection equipment manufacturers, the FAA should develop and implement a plan to collect specific data on false-alarm rates and operator alarm resolutions. The FAA should also develop a plan to collect operational data on detection rates, with and without operator involvement. In addition, the FAA should ensure that the data-collection plan is carried out and systematically documented.

The main conclusion in the final report of the FAA Operational Demonstration Project at San Francisco International Airport was that the time required to resolve alarms must be reduced to increase throughput. However, the panel concluded that more data are necessary to evaluate the combined performance capability of deployed equipment and

BOX 6-1
FAA Conditions for the Use of Explosives-Detection Equipment

- The equipment shall be used continually during periods of passenger baggage acceptance.
- Air carriers are encouraged to allow bags from other air carriers to be screened with the equipment to maximize its use.
- The equipment shall be available on demand to aid in threat resolution for bags that may have been identified as suspicious through other screening processes.

TABLE 6-3 Summary of Open Testing of CTX-5000 SP at San Francisco International Airport

Test	Sample size	P_d (% X)	$P_{fa}{}^a$ (% Y)
Machine (automated)	131	102	150
Machine + operator	131	89	5

[a] False-alarm rate was for regular passenger baggage.

operators. Data on the reliability, maintainability, and availability of the deployed equipment should be collected and made available to the airlines.

Recommendation. Controlled testing, such as the testing done during the Airport Demonstration Project, should be conducted for a variety of equipment configurations. Data should be collected and maintained on performance (e.g., probability of detection) and the test conditions. The reports on the Manila airport demonstration projects should be completed and reviewed for confirmation of or challenges to the results of the San Francisco tests.

Testing

Certification tests only reflect the ability of the equipment to identify a bag that contains an explosive. The detection rate is based on the alarm being set off for a bag containing the explosive, even if the alarm was triggered by a nonexplosive object in the bag. Certification testing does not measure alarm resolution and does not include testing in the operational environment of an airport. In the panel's opinion, some of the problems encountered with the CTX-5000 SP in the field can be reasonably attributed to the limitations of certification testing. Furthermore, under current certification guidelines, equipment certified in the future may encounter similar problems.

Recommendation. During certification testing, the FAA should, whenever possible, measure both true detection rates (i.e., identification of the correct location of an explosive when an alarm occurs) and false-detection rates (i.e., an alarm set off by something in a bag other than an explosive).

Recommendation. The FAA should assess the feasibility of including airport testing of an explosives-detection system as part of the certification process.

Recommendation. The FAA should include the ability of explosives-detection equipment to aid the operator in resolving alarms as part of certification testing. For example, alarm resolution should be a factor in the determination of throughput rate, detection rate, and false alarm rate.

The bag set used to determine the false-alarm rate during certification testing does not contain many of the items normally found in passenger bags, such as foods and liquids. This difference could account for the 50 percent increase in false-alarm rates in the field over the certification standard. Furthermore, the throughput rate measured during certification testing is based on the continuous flow of bags and does not include time for alarm resolution. Consequently, the throughput rates measured during certification testing are not representative of throughput rates in the field.

Recommendation. The bag sets used for estimating false-alarm rates during certification testing should include all of the items usually found in checked passenger bags (e.g., sand, books, food items, liquids, jewelry, toiletries, and clothes).

7

Trace Explosives Detection

Trace explosives-detection devices (TEDDs) represent a significant portion of the current FAA/SEIPT deployment of explosives-detection equipment. Configured first as vapor detectors, and more recently as trace (e.g., minute particulates) detectors, these relatively low cost, portable, sensitive, chemically specific devices are sometimes used for screening carry-on baggage. TEDDs are being used in the current deployment for the resolution of alarms of checked baggage from bulk explosives-detection equipment and for checking suspicious carry-on hand baggage.

The detection of trace amounts of explosive does not necessarily reveal the presence of a bomb. An alarm from a TEDD signifies the presence of vapors from or particles of explosive material that can only be connected inferentially to the presence of an explosive. Although TEDDs can be used to determine the type of explosive material, they cannot be—and are not—used to determine the quantity of explosive material. Furthermore, trace-detection techniques may be vulnerable to unsophisticated countermeasures, and the absence of trace amounts of an explosive does not guarantee that no explosive is concealed in the bag.

The purchase and deployment of TEDDs in the United States was mandated before a TEDDs-certification program was developed on the assumption that their deployment would improve aviation security. Since then, at least one TEDD test and evaluation plan has been drafted by the FAA that includes an evaluation of the detection of traces of explosive materials on carry-on passenger baggage by TEDDs and associated procedures (DOT, 1997). The plan is intended to establish a long-term method of collecting operational data for the purpose of assessing the reliability, performance, and costs associated with deployed TEDDs.

Unfortunately, standards and procedures for evaluating the performance of TEDDs and TEDD operators were not available at the time of this writing. In this chapter, the panel introduces the principles of trace detection, discusses the current deployment of TEDDs, outlines a short-term test and evaluation program, and suggests a course of action for developing a certification process for TEDDs and TEDD operators.

TABLE 7-1 Most Effective Techniques for Sampling Explosives for TEDDs

Explosive	Technique
RDX	Wiping
PETN	Wiping
TNT	Wiping
EGDN	Vapor detection
OMNT	Vapor detection
PMNT	Vapor detection
DMNB	Vapor detection
NG	None known

PRINCIPLES OF TRACE DETECTION

The detection of trace amounts of explosives on suspect articles is done in steps (Figure 7-1). First the suspect article is wiped with a sampler,[1] which is either a patch of material or a vacuuming device that draws air through the material for concentration of the sample (Table 7-1). For example, an operator might wipe the handle of a suitcase, the outside of a case, a computer switch, or a suspected area with a sampler. The sampler is then manually put into the TEDD where the collected material is thermally desorbed from the wiping substrate. The vapor is then transferred to a detection system in the device for chemical and quantitative characterization.

[1] Wiping is not the best method of sampling for all explosives. The optimum sampling technique varies by explosive type.

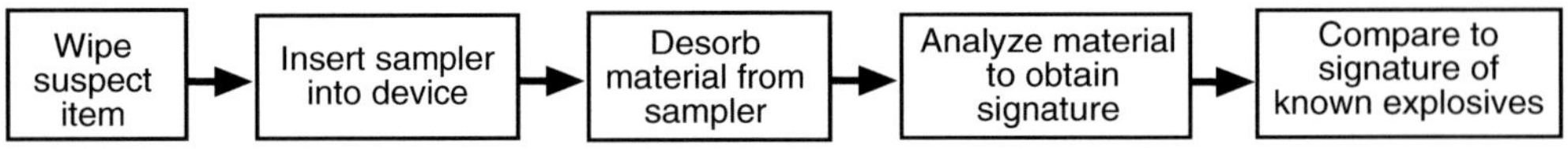

FIGURE 7-1 Operational steps of a trace explosives-detection device.

The detection system separates chemical species and creates a signature characteristic of the chemicals present. The signature is then compared to the signature of actual explosive materials, and if the combination meets the trace alarm threat criteria, the operator is notified of an alarm.

Several types of physical and chemical techniques are used in the TEDD to identify explosive materials (Box 7-1). The rationale for using TEDDs is only valid if it is assumed that the bomb maker will contaminate his or her hands with explosives and subsequently contaminate the bomb container (e.g., a radio) during fabrication of the explosive device.

Research conducted on the fabrication of explosive devices under controlled conditions has shown that it is very difficult, but not impossible, to avoid contaminating the surface of the primary container housing the explosive device. Fingerprints with sufficient material to be detected by commercial TEDDs are usually left. Plastic explosive materials tend to be "sticky" and are transferred to anything that is touched. Tests using C4 explosives as a model material suggest that, even with only 10 percent yield (i.e., the sampler only collects 10 percent of the material available for sampling), enough material is left from tenth-generation fingerprints to be detected by current TEDDs (Greshem et al., 1994). It should be noted, however, that the sampling yield is a function of the surface sampled (e.g., telephone, radio, computer), the sampling technique, and the sampler material. The ease of removing particles from a surface depends on several factors, such as the amount of explosive present, molecular and electrostatic attraction between the particle and the surface, adsorbed surface films, particle size and shape, degree of particle agglomeration, surface roughness, and the duration of contact. For a TEDD to be effective, the surface of interest must be adequately sampled, and the TEDD must be operating within known limits of detection and chemical selectivity.

DEPLOYMENT

As a part of the congressional mandate to deploy explosives-detection equipment, the SEIPT has scheduled the deployment of 631 TEDDs operated by 30 carriers at a cost of approximately $100,000 per deployed device. As of January 31, 1999, 366 TEDDs had been deployed in 39 airports, including all 19 Category X airports and one-third of Category I airports. The TEDDs are being used to resolve alarms of checked baggage screened by bulk explosives-detection equipment and to screen electronic devices and other carry-on items not cleared by x-ray at airport security checkpoints. The status of the current deployment is shown in Table 7-2.

The deployed TEDDs are being evaluated separately and not as part of an overall system, such as the TAAS. A study of how TEDDs are being used in airports would provide a much better picture of their utility as part of an overall security system, but the operational performance of TEDDs cannot be evaluated because the detection sensitivity and

BOX 7-1
Operating Principles of Chemical-Analysis Techniques Applied to Trace Explosives Detection

Ion-mobility spectrometry. Separation of electronegative molecules where lower-mass molecules move faster than heavier ones. Discrimination is based primarily on molecular weight. Detection is accomplished by negative-ion current collection.

Gas chromatography. Partitioning of molecules by solubility between a liquid phase and a gas phase in a long tube to separate them in time. Total time indicates molecular characteristics. Detection is accomplished by a wide variety of methods, including chemiluminescence and negative-ion current collection.

Chemiluminescence. Nitrogen-containing compounds are detected by measuring the light emitted from excited states.

Atmospheric pressure electron ionization. Explosive molecules are charged selectively for molecular analysis by ion-mobility spectrometry.

TABLE 7-2 Status of TEDD Deployment (as of January 31, 1999)

Manufacturer	Model	Number Deployed	Number to Be Deployed in FY 1999	Total
Barringer	Ionscan 400	178	220	398
Ion Track	Itemizer	80		80
Thermedics	Egis 3000	55		55
	Egis 2	—	5	5
IDS	Orion	53		53
To be determined			40	40
Total		366	265	631

chemical selectivity of the installed units were not previously determined in the laboratory. Because the performance capabilities of the TEDDs are not known, operator performance cannot be measured either. Furthermore, no standards or procedures for testing and data collection have been established. Thus, P_d, P_{fa}, sampler efficiency, and operator proficiency cannot be measured.

TESTING AND EVALUATION

Quantifying the improvement in aviation security by explosives-detection equipment is difficult. The detection and interception of an explosive before it is loaded onto or carried aboard an aircraft would be a clear measure of efficacy. However, to date, no incidents of this type of detection have been recorded in the United States. In fact, the only statistics available on aircraft bombing attempts are on attempts that actually resulted in onboard explosions.[2] The number of bombings and typical baggage-flow rates suggest that there is less than one bombing attempt against U.S. aircraft every 10 years. Thus, only one bag out of several billion contains an explosive. Therefore, it is almost impossible to measure directly the efficacy of TEDDs in preventing actual bombing attempts, and their effectiveness can only be estimated through comprehensive testing and evaluation, which should include the steps shown in Figure 7-2. Ultimately, the improvement afforded to the overall system by TEDDs can be evaluated by their impact on the SEF.

The most practical way to assess the contributions of TEDDs to overall aviation security is to establish performance requirements. Until a certification standard is established, the FAA could initiate a short-term (e.g., six-month) program to assess currently deployed TEDDs on the basis of their operational effectiveness. The data from operational testing could then be used to assess their effect on overall security (e.g., the proposed SEF).

[2] No bombings of U.S. commercial aircraft have been confirmed since the initiation of the congressionally mandated deployment of explosives-detection equipment.

Establishing the Trace Threat Amount

The amount of each threat material a TEDD must be able to detect (the "trace threat amount") must be established, not as a range but as a specific amount for each explosive and specified as a function of area because sampling is area dependent. FAA studies to establish trace threat amounts are ongoing.

Testing Equipment

Testing a TEDD involves introducing a known amount of explosive material from a sampler used by the manufacturer (e.g., a cloth wipe). The amount on the surface of the sampler—which is then introduced into the detector—must be a known amount not a random amount from wiping an area on a suitcase where explosive material may or may not be present or that may be difficult to transfer to the wipe. The amount used to test the TEDD could be determined experimentally. For example, if the TEDD's sampling of a defined threat amount (e.g., 1 ng from a specific area of a contaminated article) with the vendor's sampler is determined to be 10 percent efficient, the TEDD would have to alarm on 100 picograms of explosive placed directly on the sampler and into the detector to be considered effective. This test would determine the ability of the TEDD to detect explosives and the utility of the sampler for transferring the explosive to the detector. This standard could then be defined as a TEDD calibration standard (TEDDCS) and could be shared with the TEDD vendors.

The TEDDCS should consist of explosive material *dissolved* in a volatile solvent to produce a known concentration prepared under the direction and control of the FAA, not the vendor. Currently, TEDDs are tested on a daily basis by vendor-supplied standards, some of which are not even

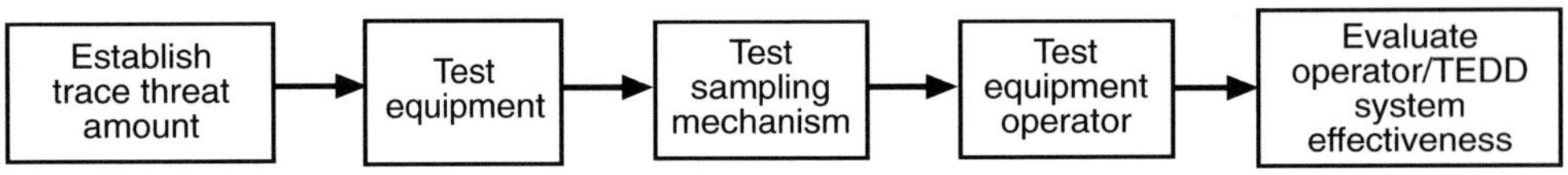

FIGURE 7-2 A process for measuring the effectiveness of the operator/TEDD system.

quantitative. The panel was informed that the FAA is preparing standards, but it is not clear whether the standards are slurries or explosive materials dissolved in a volatile solvent (suspension solutions) (Fox, 1998). Slurries or suspension solutions pose significant difficulties for establishing known quantities and should not be used for this purpose (INEL, 1998).

Testing the Sampling Mechanism

Testing the sampler will involve defining a standard surface or surfaces (e.g., a suitcase handle) and standard (known) contamination level (e.g., the amount left from a tenth-generation fingerprint) and measuring how well a sampler collects a sample from a known amount of explosive material deposited on a known area (e.g., of fingerprint size). No test protocol has yet been developed to determine the efficacy of a sampler.

Testing the Equipment Operator

Once it can be shown that a TEDD sampler can collect an adequate amount by "wiping" a known area in an operationally compatible mode, the operator can then be tested. Unlike the automated EDS called for in the Aviation Security Improvement Act of 1990 (PL 101-604), TEDDs will only work if the operator samples a surface correctly. Given their dependence on operator performance, it is incumbent upon the FAA to ensure that TEDD operators sample baggage and other objects properly. This will probably require a protocol for certifying TEDD operators, as well as a protocol for maintaining their performance at a certified level.

Measuring the performance of a TEDD operator can be done using blind and double-blind tests, which require test objects that will not be recognizable as test objects by the operator. If, for example, an operator is handed vinyl luggage handles, floppy discs, or zippers (the current test items for experimental verification and validation of TEDDs), the operator will probably recognize them as test objects. Thus, this procedure will not serve as a blind test. Furthermore, there is evidence that the current test suspension is visible when deposited on test substrates (INEL, 1997). For blind tests, normal hand-carried items on which the trace threat has been deposited in a concealed manner must be used. So far, no procedures have been developed for placing a TEDDCS on realistic bags, for performing blind and double-blind tests with such objects, or for systematically evaluating TEDD operators in an airport environment. Blind testing is essential to measuring the performance of the operator-TEDD combination.

Evaluating False Positives

False positives and their impact on aviation security and airline/airport operations were not evaluated in this report. The P_{fa} of deployed TEDDs has been reported to be on the order of about 2 percent (INEL, 1997). Because the sensitivity of TEDDs (i.e., the amount of explosive required to set off an alarm) is not known, identifying the source of false alarms is problematic. Once the devices are operating at a known and specified detection level, the causes of false alarms can be identified. That is, the cause of all cleared (false) alarms should be documented so that a list of interferants can be tabulated. These data would be invaluable for future development and deployments.

Evaluating False Negatives

False negatives are exceedingly difficult to measure and assess. A false negative could be the result of a TEDD being out of specification or an operator not following proper sampling procedures, or it could be the result of a very careful bomb maker. In other words, a very "clean" bomb maker who uses a very good concealment technique could conceivably defeat a TEDD even if it is operating within specifications and the operator is performing flawlessly. Thus, false negatives are partly dependent on the terrorist and cannot be completely eliminated by improvements in TEDD technology or operational protocols.

CONCLUSIONS AND RECOMMENDATIONS

Trace explosives-detection techniques detect traces of explosive materials, which may or may not indicate the presence of a concealed explosive. The best way to ensure that TEDDs are contributing to the security of commercial aviation is to ensure that they meet yet-to-be-defined FAA certification requirements. However, because sampling is a dominant factor in determining the effectiveness of the technique, separate certification and operational testing will be necessary for measuring the performance of equipment operators. The certification of TEDDs will be necessarily independent of the certification of TEDD operators to ensure that both are working properly. In the panel's opinion, the FAA has not provided a viable quantitative primary standard, and the current method of deposition on test objects does not mimic fingerprints.

Recommendation. The FAA should establish separate certification protocols for trace explosives-detection devices (TEDDs) and TEDD operators.

Recommendation. The FAA should establish a specific threat trace amount that trace explosives-detection devices must be able to detect to be certified.

Recommendation. The FAA should develop a trace explosives-detection device (TEDD) calibration standard (TEDDCS)—based on an established threat trace amount—for measuring the capability of a TEDD to detect explosives introduced directly onto the sampler and into the instrument. This standard should be used for certification testing, as well as for performance verification in the field. The TEDDCS should also be made available to TEDD manufacturers for developmental purposes.

Recommendation. The trace explosives-detection device calibration standard should be dissolved in a volatile solvent to produce a known concentration. Slurries or suspension solutions should not be used for this purpose.

Recommendation. The FAA should develop a test to measure the trace explosives-detection devices sampling mechanism (sampler). The test should measure how well a sampler collects a known amount (trace explosives-detection device sampling standard) of explosive material deposited on a known area in a manner consistent with the trace explosives-detection device's operational protocol.

Recommendation. The FAA should develop a certification test for operators of trace explosives-detection devices that measures their ability to screen baggage and other objects. In addition, the FAA should periodically conduct blind and double-blind tests to monitor the performance of certified operators.

Recommendation. The panel recommends that the FAA continue to monitor false alarms of deployed trace explosives-detection devices (TEDDs). The false alarms should be documented, including the causes, and a list of interferants compiled to be used for the future development and deployment of TEDDs.

8

Computer-Assisted Passenger Screening and Positive Passenger-Bag Matching

Every year more than 600 million passengers travel by U.S. air carriers. On average, each passenger checks 0.8 to 1.8 bags, depending on the characteristics of the passenger (domestic, international, business, leisure). In addition, passengers have (on average) 1.0 (leisure) and 1.5 (business) carry-on bags. One billion passenger bags per annum is not an unreasonable estimate. Assuming that the threat of an onboard explosion destroying a U.S. commercial aircraft remains relatively constant at one every 10 years, the detection of explosive devices in checked or carry-on baggage is a daunting task. Neither examining every bag physically nor screening every checked bag with deployed EDS technology is currently feasible.

POSITIVE PASSENGER-BAG MATCHING

One security procedure adopted for international flights is to match passengers with their checked baggage prior to the departure of the flight, commonly referred to as PPBM (positive passenger-bag matching). PPBM involves off-loading a passenger's checked baggage from the airplane if the passenger does not appear at the gate for departure. The rationale for this procedure is that a passenger will not wittingly fly on an aircraft that will be destroyed by a bomb during flight. However, this security procedure will not work for a suicide bomber or for a passenger who is unaware that he is carrying an explosive device in his baggage (a so-called "dupe").

It has been suggested that PPBM be used on U.S. domestic flights. The flight and baggage statistics, however, as well as the current design of the U.S. domestic airline system—which relies on major hubs for efficiency and economy—indicate that 100 percent PPBM for domestic flights would not be practical because it would probably reduce the efficiency and increase the cost of commercial flights.

PPBM is now used effectively for passengers on single-leg flights, but baggage matching is not maintained for multileg flights. In other words, PPBM is not repeated after the first leg of a multileg flight, either for intraairline or interairline flights. Therefore, the security afforded by PPBM is compromised for any flight that receives checked baggage from another flight. Furthermore, passengers who check their luggage at the gate may not be subject to PPBM.

COMPUTER-ASSISTED PASSENGER SCREENING

The report of the White House Commission on Aviation Safety and Security (1997) indicated that profiling is just one component of a comprehensive, layered security program, and, like other security measures, profiling may become less necessary as more efficient screening technologies are introduced. Based on readily available information, passengers can be separated into a very large majority of people about whom enough is known to conclude that they present little or no risk and a small minority who merit additional attention. The commission report endorsed efforts by airlines and the FAA to develop a profiling system.

For many years, law enforcement organizations in the United States and elsewhere have used profiling systems to identify individuals who might be involved in illegal activities. The U.S. Customs Service, for example, uses a profiling system to separate known narcotics traffickers (high-risk individuals) from low-risk individuals who enter the United States via airlines, ships, or other modes of entry. U.S. Customs agents only search the bags of passengers arriving from foreign destinations who fit a known profile, thus focusing available manpower on the individuals most likely to be carrying drugs or other illegal items into the United States. The objectives of profiling are effectiveness, efficiency, practicality, and economy of resources.

The system currently used by the FAA and the airlines, called CAPS (computer-assisted passenger screening), is an automated procedure that reviews data in airline passenger records and matches those data to criteria developed by the FAA Office of Civil Aviation Security Intelligence based on the characteristics of terrorists who have perpetrated attacks against commercial aviation. CAPS separates airline passengers into two groups, high risk and low risk. The carry-on and checked bags of high-risk passengers (called *selectees* in airline security parlance) are subjected to additional security measures, including either PPBM or screening by an FAA-certified EDS.

All U.S. air carriers were required to implement and use CAPS by December 31, 1998. Airline representatives have indicated that the implementation of CAPS has not created unreasonable difficulties and that CAPS is superior to a manual passenger profiling and tracking system, although most airline officials also believe it is too soon to judge the effectiveness of CAPS. The number of CAPS selectees varies by airline, route, and location but averages from 3 to 8 percent of all passengers (Padgett, 1998). Although CAPS appears to be an effective component of aviation security and is anticipated to be an important part of a TAAS for the foreseeable future, a quantitative means of evaluating its effectiveness has not been implemented.

The issue of civil liberties was raised in the White House Commission Report, which concluded that fundamental civil liberties should not, and need not, be compromised by a profiling system. The commission established a Civil Liberties Advisory Panel to investigate the issue and recommend ways to safeguard civil liberties (White House Commission on Aviation Safety and Security, 1997). The U.S. Department of Transportation also submitted the profiling elements of CAPS to the Civil Rights Division of the U.S. Department of Justice for review, which determined that CAPS did not violate the constitutional prohibitions against unreasonable searches and seizures and did not discriminate on the basis of color, gender, religion, or ethnic origin. The Department of Justice recommended, however, that the FAA periodically review the CAPS program to ensure that it adheres to constitutional requirements and civil rights laws, that airlines seek governmental approval prior to altering the program, and that employees responsible for the operation of CAPS be trained in civil liberties.

CONCLUSIONS AND RECOMMENDATIONS

The 1997 White House Commission on Aviation Safety and Security recommended that the FAA implement full PPBM. According to the FAA's interpretation of this recommendation, full PPBM involves matching passengers, either randomly selected or identified through CAPS with their bags (GAO, 1998). The panel concluded that 100 percent baggage matching for domestic flights is not practical and that the combination of PPBM with CAPS—when fully implemented—will improve aviation security further. However, it is not clear that PPBM is continued when a selectee changes planes at a connecting airport or disembarks during a stopover. The panel believes that selectees at originating airports should be considered selectees, and therefore subject to PPBM, on subsequent connections.

CAPS appears to be effective for identifying selectees for further security measures, such as bag matching or bag screening. The panel concluded that CAPS effectively focuses personnel and equipment on high-risk passengers and does not impede the efficiency or productivity of air travel. Nevertheless, the panel believes a means of quantitatively characterizing the effectiveness of CAPS should be developed to ensure its long-term viability.

Recommendation. The FAA and the airlines should extend the computer-assisted passenger screening program to include interairline coordination (sharing of data for connecting passengers) and interfaces with other components of the overall security system (e.g., explosives-detection systems, positive passenger-bag matching).

Recommendation. The FAA should develop a quantitative measure of the performance of computer-assisted passenger screening.

Recommendation. Passengers designated as selectees at the origination of their flights should be considered selectees for all connecting (or continuation) legs of their flights.

9

Human Factors

Human factors should be an important consideration in the deployment of EDSs, noncertified bulk explosives-detection equipment, and TEDDs. The primary performance measures for all of these systems are throughput and operational errors, which are often considered to be human errors. Even if the equipment itself fails, an analysis usually reveals an underlying human error in management, installation, or maintenance. For this reason, the performance of equipment in the laboratory should not be considered an estimate of system performance but rather an upper bound. In congressional testimony by the U.S. Department of Transportation Inspector General's Office, human factors were acknowledged to be crucial to system effectiveness (DOT, 1998). A telling example (detailed in Chapter 10) is the sensitivity of overall system effectiveness to human performance in resolving alarms of the CTX-5000 SP. Human performance can be improved by improving training or by changing the equipment and design and operating procedures.

Human performance, as measured by P_d and P_{fa}, varies with the configuration and use of the EDS. For example, P_d and P_{fa} of the human/EDS system are sensitive to throughput rate, which is highly dependent on how bags are selected for screening. Thus, human factors should be explicitly considered throughout the design and deployment process rather than treated as a late addition. Inherent deficiencies in a particular system cannot be remedied by *post facto* operator training.

MODELS OF BULK AND TRACE SCREENING

Although both bulk and trace explosives-detection equipment are based on different technologies, have different human interfaces, and are deployed in different branches of the TAAS, they are both inspection systems intended to detect threats with minimal false alarms and maximum throughput. Currently, a common inspection model is used to evaluate the deployment of both kinds of systems. This model (Drury, 1989), which was derived from earlier models (Harris and Chaney, 1969; Sinclair and Drury, 1979), has been used for some time for industrial inspections and aviation structural inspections. In Figure 9-1, the functions are defined in generic terms along the center column (i.e., without reference to a specific hardware or application). Each function can be assigned either to a human operator, to a machine, or (occasionally) to parallel human and machine systems. In the deployment of bulk and trace systems, the allocation of function is part of the system development process. Thus, deployment involves managing installation and operation rather than reallocating functions. Operational data may reveal, however, that functions were not allocated optimally or that the implementation of particular functions could be improved. The side columns of Figure 9-1 show whether a generic function has been allocated to a *human* or a *machine*. Note the distinct differences between the bulk and trace systems, which require different capabilities by the human operators.

Figure 9-1 shows only one component in the overall TAAS. This component is linked to other components by how bags arrive for screening and by the alarm-resolution procedures. For example, if a TAAS configuration has two EDSs in series with alarms on the first EDS resolved by the second, the human decision in the first EDS may not be critical. If noncertified bulk explosives-detection equipment is used that does not have image display capability but uses a go/no-go indicator, the operator is not involved in determining if there is an alarm but is involved in resolving it.

FACTORS THAT AFFECT HUMAN AND SYSTEM PERFORMANCE

In Table 9-1, the factors that affect human and system performance are broken down by task, operator, machine, and environment. In bulk systems, the visual search function is allocated to machine hardware, a reasonable first step because the human visual search function is uniformly unreliable (Hou et al., 1994). In TEDDs, however, the search

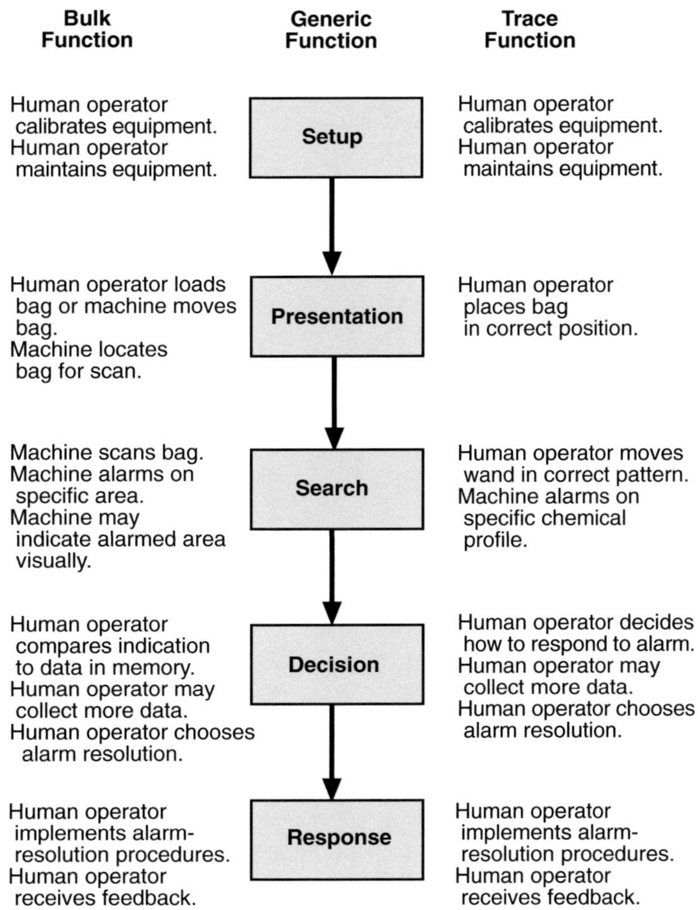

FIGURE 9-1 Role of the human operator in explosives detection.

function involves an operator wiping or passing a wand over the bags, a human function of uncertain reliability.

As Table 9-1 also shows, many management support issues affect the operator and, hence, the performance of security equipment. First, management must actively ensure that procedures are followed (e.g., maintenance, alarm resolution) and that decisions are made (and actions taken) based on security concerns. The clearance of an alarmed bag represents a potentially dangerous situation because explosives are more likely to be present in alarmed bags than in unalarmed bags. Therefore, clearing or not clearing a bag can be a perceptually difficult and emotionally charged decision. The operator requires management backup in these decisions and in the application of resolution procedures. Second, management is responsible for system-wide conditions that affect operator performance (e.g., number and selection of bags reaching the operator, time pressures on the operator, training, and feedback to the operator). Third, management is responsible for job design, which can affect, for example, operator turnover. A well designed job affords operators great latitude in making decisions (i.e., control of their own work), low levels of psychological demand (i.e., workload and work difficulty), and high levels of support (i.e., from supervisors and peers) (Karasek and Theorel, 1990). Traditional x-ray screening jobs have few of these desirable characteristics and, in addition, have socially undesirable hours and low pay. Therefore, high levels of turnover are to be expected and have been noted in previous NRC reports (1996a, 1997). Job design is important for bulk and trace explosives-detection equipment. Because additional operator training is required (particularly for bulk EDS), management has both an incentive and an opportunity to rethink job designs to reduce the turnover of trained operators.

Table 9-1 also shows operator/machine interfaces and personnel issues, such as operator abilities (determined by selection, retention, or job rotation) and operator training (quality and timeliness). For bulk equipment, such as the CTX-5000 SP, the human/machine interface is a manipulable screen image, usually in false color, with the suspect area highlighted. The operator may have to access other

TABLE 9-1 Factors That Affect Operator and System Performance

Function	Task	Operator	Machine	Environment
Setup	1. Calibration procedures 2. Maintenance procedures 3. Time available for tasks	1. Ability to follow procedures accurately 2. Training (quality and timeliness) for calibration and maintenance	1. Operator interface for calibration 2. Operator interface for maintenance	1. Management support of equipment calibration and maintenance 2. Location of equipment 3. Availability of equipment operating procedures and other job aids
Presentation	1. Baggage selection	1. Training	1. Physical layout of equipment	1. Management control over baggage selected 2. Perceived and actual passenger pressure
Search[a]	1. Defined wand search pattern	1. Training (quality and timeliness) for search pattern 2. Operator dexterity	1. Human interface (e.g., wand, alarm indicator)	1. Management support to ensure procedures are followed 2. Perceived and actual passenger pressure
Decision	1. P_d 2. P_{fa} 3. Time available	1. Knowledge of threats 2. Knowledge of potential false-alarm items 3. Experience, overall and recent 4. Ability to make a decision 5. Resistance to passenger pressure	1. Operator interface to display design 2. Operator interface to acquire additional information	1. Management support for following correct decision procedures 2. Perceived and actual passenger pressure
Response	1. Alarm-resolution procedures 2. Management provision of feedback	1. Knowledge of alarm-resolution procedures 2. Resistance to passenger pressure	1. Interface to other systems	1. Management support for alarm-resolution procedures 2. Perceived and actual passenger pressure

[a] Note that search has been largely allocated to machine functions for bulk explosives detection.

views of the highlighted area to determine its spatial surroundings or even call for more data, such as additional scans. Even if operators can confidently classify the alarmed area as a false alarm, they must perform a visual search (which has potentially limited reliability) of the rest of the image because a known false alarm in a bag could be a diversionary tactic to draw attention away from a smaller true threat in the same bag.

The operator/machine interfaces in TEDDs, which are capable of determining specific chemical species, are simpler because chemical false alarms are less frequent. Thus, the interface can be as simple as a two-stage indicator (alarm/no alarm). In TEDDs, the interface is not specifically designed to assist the operator in resolving alarms, and alarm resolutions are typically procedural (e.g., eliciting information from the passenger whose bag caused the alarm).

DEPLOYMENT ISSUES

Now that explosives-detection equipment has been (and continues to be) deployed, concerns have arisen about physical installation, the training of security personnel, and the integration of the equipment into ongoing security and baggage-handling operations. One particular concern is that the CTX-5000 systems in many locations are operating at a fraction of their rated capacity (DOT, 1998). This underutilization may be attributable partly to alarm resolution time, partly to location, and partly to the number of bags being sent to them. Underutilization poses a potential problem for the maintenance of operator skills, particularly the skills required for resolving alarms, because underpracticed skills often deteriorate. At some locations, the throughput rate has been so low that operators could even lose their skills for operating the equipment. Reliable data on improvements or the deterioration of screening skills as a function of experience are sparse, but in other disciplines skill maintenance has been a problem.

In analogous military systems with rare threats, the military's solution has been to increase training, particularly embedded training (i.e., the introduction of simulated threats into the system during operations) (Walsh and Yee, 1990). Responses to these threats can be quickly analyzed and feedback provided to the operator. A technology that could be used for embedded training in airport security screening (called threat image projection system [TIPS]) has been developed but is not being used on all conventional x-ray scanners or bulk explosives-detection systems. With an in-service measurement system, such as TIPS, it should be possible to establish a performance timeline, including an initial learning curve, long-term fall-off in performance, and the benefits of recurrent training.

The deployment of TEDDs has led to some integration problems (e.g., calibration and maintenance) but fewer than have been encountered for bulk explosives-detection equipment. The FAA has not provided realistic, precisely specified random challenges to TEDDs for testing and feedback, although they have been developed for other equipment used at security checkpoints and the checked-baggage stream (e.g., FAA test objects, TIPS). Currently, performance measurement for TEDDs is not well coordinated, and feedback is dependent on the items presented by passengers and the reliability of alarm resolutions.

One problem common to the development and deployment of both bulk and trace explosives-detection equipment is lack of human-factors support. Not many human-factors engineers are familiar with security issues, and human-factors engineers from other domains have not received enough training to be helpful. Unfortunately, although the FAA has expanded its efforts, its resources are limited.

A number of human-factors issues discussed in this chapter have become apparent only as explosives-detection equipment has been deployed. Although none of these issues is new, at least to human-factors professionals, the fact that they are being raised as primary concerns now shows that they were not addressed during the research and development phase that proceeded this deployment (DOT, 1998).

The most serious issue is that the FAA certifies equipment only and not the human/machine system. Thus, certification provides only a bound on total system performance, and when the system is deployed, the performance level of the equipment/operator system is sometimes well below the performance level of the equipment alone. The FAA should continue to certify equipment but should also certify that the human/equipment system meets performance requirements. Thus, if the operator must resolve each alarm, the operator should have demonstrated that he or she can correctly identify threats and correctly clear nonthreats with defined probabilities. If equipment manufacturers know that systems will be tested as a whole, they will have to consider the human role in the system and include design displays and procedures to support it:

- Bulk explosives-detection equipment must provide support for alarm resolution. The CTX-5000 produces a visual display, but whether this is the optimum way to support alarm resolution has not been determined.
- The trace explosives-detection equipment must be easy to maintain and calibrate. The contribution of TEDDs to overall system performance (i.e., TAAS) cannot be assessed unless the equipment operates with known detection parameters.

Management support is essential for many operator functions. Responsive management practices (e.g., setting and maintaining policies that emphasize effectiveness rather than throughput), appropriate job aids (e.g., devices for calibrating and maintaining trace equipment), and the time and training to use them all affect operator performance. A job with a very low ratio of job-tenure time to skill-acquisition time raises questions about job design and management commitment. The addition of new technology that is not well integrated into the overall security system only increases the demands on equipment operators.

Timely, controlled feedback is essential to the development and maintenance of high performance levels and is essential to embedded training. If the only feedback operators receive is the outcome of imperfect alarm-resolution procedures, managers will not have control. Effective feedback requires systematically challenging the system with well characterized simulated threats. Current simulated threats are FAA test objects and modular bomb test sets (MBTSs). For the current x-ray CTX-5000 equipment, TIPS can be used, although it may not work on all current x-ray systems. Furthermore, an equivalent system to TIPS will have to be developed for trace equipment. On a larger scale, challenges can be provided by one-time evaluations, such as double-blind testing,[1] but these are more useful for audits than for regular feedback because of their irregular use.

The FAA test objects are currently used to evaluate the performance of airport security systems on a regular basis, with specified procedures for dealing with detection failures. These tests should be expanded and revised for both current systems and newly deployed technologies. Overall performance could be improved by testing with more realistic threat objects.

Once a performance measurement system is in place, the FAA should consider certifying individual equipment operators. The FAA could issue performance standards and conduct evaluations or examinations, as it does for other occupations in civil aviation, such as pilots or maintenance technicians. However, the responsibility for training operators to FAA standards should rest with private industry, such as airlines, security companies, and trade schools. The introduction of new equipment, which increases the complexity of the job, could then provide for job progression, which—coupled with redesigned jobs and operator licensing—could increase job tenure and reduce turnover. The annual turnover rate is more than 100 percent in many locations, and with longer training times (for CTX operators), the personnel problem is becoming more acute. The measures suggested here could lower turnover rates by addressing underlying reasons for worker turnover.

Current evaluations of both older screening systems and the newly deployed systems are based on direct performance measures, including missed detections, false alarms, and response times. However, measuring performance does not in itself explain why that performance occurred or how it could be improved. For example, if a bulk system alarm is resolved improperly by an operator and results in a missed detection, the reason the error occurred must be determined. Possible causes could include a poor display for the operator, the assumption that an alarm is triggered by a familiar "recognized" material, failure to obtain the correct additional information, time pressure, and perceived customer/management pressure. In any particular case, the causal factors usually lead back to underlying latent failures, such as inadequate training, poor management, or even improper allocation of function in the design (Reason, 1990).

Detailed error analyses will increase the cost of data collection and will require at least a function-level model of the screening process (Neiderman and Fobes, 1997). However, the data would provide much needed guidance for raising performance levels. The same analyses could be used to determine the reasons for successful (or correct) decisions by equipment operators.

[1] An example of double-blind testing is an FAA employee posing as a terrorist trying to sneak a simulated bomb onto an airplane without the knowledge of the security equipment operators.

CONCLUSIONS AND RECOMMENDATIONS

Human operators are integral to the performance of all deployed explosives-detection equipment. Because fully automated explosives-detection equipment is not likely to be developed in the foreseeable future—particularly with respect to alarm resolution—human operators will continue to be immensely important to realizing the potential of deployed security hardware. Current certification testing of explosives-detection equipment, however, only defines the operational capability (or performance) of the equipment, and human factors have resulted in a lower operational performance level than the certified detection capability of the equipment. Thus, the human operator/equipment combination should also be required to meet performance requirements for FAA certification. Furthermore, standards of operator performance for all systems should be raised and then monitored and the results regularly provided to airlines.

Recommendation. In addition to certifying explosives-detection systems, the FAA should ensure that these systems can be operated (by a human operator) at a specified probability of detection, probability of false alarm, and throughput rate.

Recommendation. The FAA should deploy more human-factors experts throughout the system. Human-factors training should be provided for at least some system managers and operators.

Recommendation. The FAA should initiate a program to improve operator performance that includes the following elements:

- measurements of the performance of the equipment/operator combination
- valid and reliable challenges (tests) for system components
- embedded training to improve and maintain operator performance
- improved test and qualification procedures for operators

10

Evaluation of Architectures

Keeping bombs off of aircraft is the primary measure of the performance of a TAAS (total architecture for aviation security). Improved security components recently deployed to minimize the probability that a bomb can be placed on an aircraft by a terrorist include CAPS (computer-assisted passenger screening), PPBM (positive passenger-bag matching), HULDs (hardened unit-loading devices), TEDDs (trace explosives-detection devices), noncertified bulk explosives-detection equipment, and EDSs (FAA-certified bulk explosives-detection systems). Blind operational testing using realistic simulated bombs will be necessary to evaluate the effectiveness of the overall aviation security system. At present the FAA has only limited blind-test data on EDSs and even less data on other components of the TAAS. Test results on all TAAS components will be necessary for a systems analysis that can estimate the results of full-scale field testing of the entire TAAS. A sensitive measure of effectiveness, such as the SEF (security enhancement factor), could be used to assess improvements to TAAS components and reduce the complexity of evaluating the whole system.

SECURITY ENHANCEMENT

A good measure of the improvement from implementation of new security measures is a reduction in the number of simulated explosives brought onboard aircraft compared to the number brought onboard under a previous (or baseline) TAAS. Realistic operational testing can be used to assess how well TAAS components prevent simulated bombs from getting through security and onto aircraft. The current MBTS has sufficient dynamic range to test the bulk explosives-detection equipment components of the TAAS, but similar test articles are not available for testing TEDDs. Because TEDDs are used to resolve alarms by bulk explosives-detection equipment, evaluating TEDD performance in an airport setting is crucial. Because it may not be practical to contaminate the test sets with trace amounts of explosives, another test device must be developed.

Security Enhancement Factor

The purpose of the SEF is to develop a measure of security enhancement based on changes in the number of bombs that defeat the TAAS and are brought aboard an aircraft. The examples that follow are not representative of the actual performance of security equipment but suggest methods of analyzing performance. The SEF is defined in terms of the ratio of the number of bombs getting through a defined *baseline* TAAS to the number of bombs getting through a modified or upgraded TAAS. If there is no improvement in the performance of the TAAS, the SEF = 1. If half as many bombs get through the new TAAS, SEF = 2. If the upgraded TAAS prevents all bombs from getting through, the SEF would be infinite. For example, assume that the FAA tests a baseline TAAS with 400 simulated bombs and 100 of them defeat the TAAS and make it onboard an aircraft. If the 400 MBTSs are then put through the *improved* TAAS and only 50 get through, the SEF would be 2 (because half as many bombs defeated the *improved* TAAS). Thus, the SEF is a system-level measurement of the performance of the total TAAS.

As was discussed in Chapter 2, the baseline system in an SEF measurement will have to be redefined as new and improved equipment becomes available and as the threat evolves with time. When new equipment is deployed (perhaps to address a new threat), measuring its impact on the TAAS may involve more than measuring the SEF with this equipment "plugged in" to the existing baseline security architecture. The sequencing of TAAS components could affect the number or configuration of threats (explosives) that are screened by the new component, thus influencing the impact of the new component on the SEF. Several combinations of potential TAAS elements may have to be assessed.

Eliminating False Negatives

The critical factor for improving the SEF (and therefore aviation security) is reducing the number of false negatives

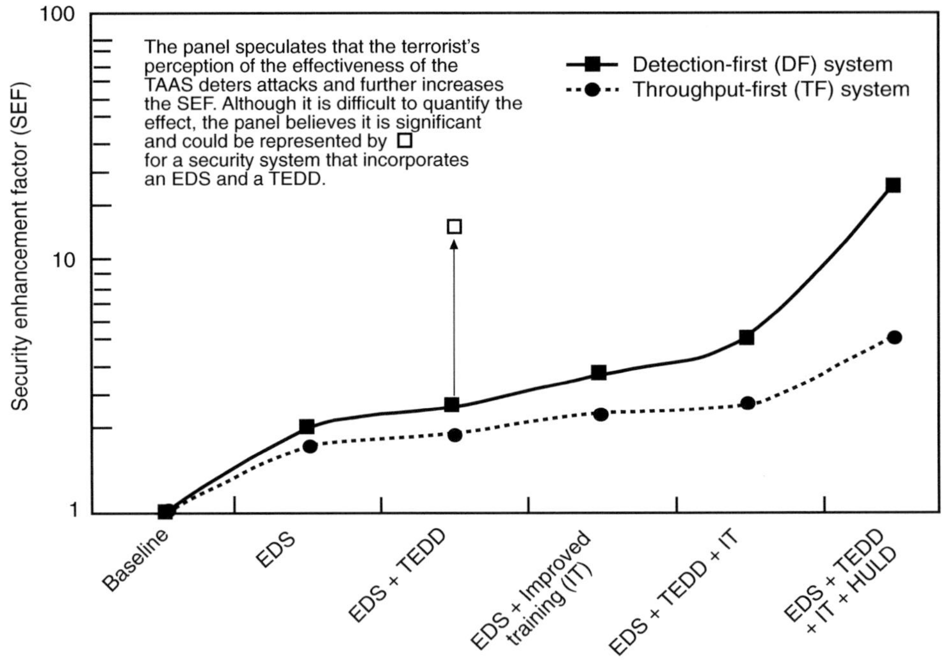

FIGURE 10-1 Comparative contributions (notional) to the SEF of detection-first and throughput-first systems.

(i.e., missed detections). Because *real* events (i.e., bombing attempts) are currently rare, the only way to ensure that the security system works is through realistic training and blind testing. If the TAAS is credible, deterrence may reduce the likelihood of a real attack. Thus, in principle, the more competent the TAAS is perceived to be in defeating bombing attempts through testing and training, the less likely a terrorist attack is to occur (see Figure 10-1).

When the P_{fa} is relatively high (e.g., higher than the airlines are willing to accept in daily operations), the large number of alarm resolutions will make it more difficult for operators to distinguish a real bomb from a false alarm. Operator decisions are currently necessarily frequent, must be made relatively quickly, and are biased toward clearing the alarm because of the infrequency of test events and the even rarer actual bomb threats. Thus, the probability of an operator missing a bomb during the alarm-resolution process is not insignificant. Improving operator performance will require substantial regular training under airport operational conditions.

Relationship between the Certification of Security Equipment and TAAS Performance

The FAA only certifies EDSs that meet their requirement for P_d, P_{fa}, and throughput rate, which are determined during certification testing with the candidate EDS in an automatic mode (i.e., without operators) at the FAA Technical Center. Under operational conditions at an airport, however, operator intervention has lowered both the P_{fa} and P_d of the FAA-certified InVision CTX-5000 SP (see Table 6-3) (FAA, 1997a, 1997b). The panel observed that operators resolve more than 30 times as many alarms for the CTX-5000 SP by using the display as they do by actually opening bags. Although the capability of the operator to identify bombs on the EDS display is not part of the current certification process, it is one of the most important aspects of reducing false alarms and false negatives. According to current certification standards, a future system with a lower spatial resolution (i.e., less capacity to resolve individual objects) could conceivably be certified in an automatic mode and yet be incapable of providing an image that could be used by an operator for resolving alarms. Therefore, the panel believes that the combined performance of the operator and equipment should also be qualified or certified in an airport environment.

ARCHITECTURES FOR AVIATION SECURITY

There are two basic philosophies for aviation security architectures. The first, the detection-first (DF) philosophy, mandates the detection of explosives at specified levels at the expense of other factors. The second, the throughput-first (TPF) philosophy, emphasizes the efficient throughput of bags through the baggage-handling system and considers

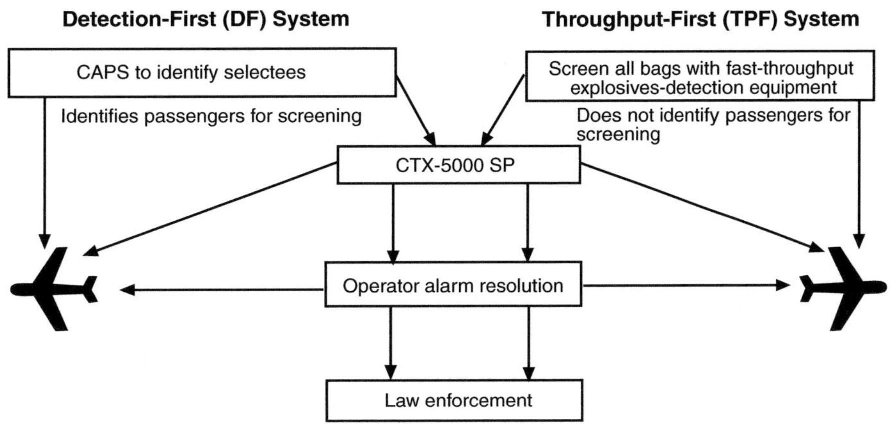

FIGURE 10-2 Schematic diagram of throughput-first and detection-first aviation security systems.

detection rates secondary. In DF security architectures, the EDS with the highest P_d is usually the first piece of equipment to screen checked baggage. In TPF architectures, the explosives-detection device with the highest throughput is placed first, sometimes at the expense of P_d. No analyses were presented to the panel of the conditions under which DF systems are more effective than TPF systems. Security systems in Europe are generally TPF systems; security systems in the United States and Israel are DF systems. The panel investigated deployed DF systems at airports in San Francisco, Los Angeles, and New York City. At JFK Terminal One, a TPF system has recently begun operations serving European air carriers. The threat vectors addressed in the two aviation security systems are shown in Figure 10-2.

The primary difference between a DF and a TPF system is the way bags are selected for more thorough scrutiny. Figure 10-3, which represents a generic checked-baggage system, shows the similarities and differences between DF and TPF systems. In the United States, the FAA has established a system for screening passengers (CAPS) that separates airline passengers into high-risk and low-risk groups. High-risk passengers (selectees) are subjected to more thorough screening. Most European airports use a TPF approach because of the difficulties of interviewing (for CAPS) the high volume of international passengers. Instead, all bags are screened, resulting in increased volume of baggage passing through security equipment (e.g., EDS) and placing a premium on throughput. Thus, as Figure 10-3 shows, in a TPF system the first level of security is the screening of all checked bags by a high-throughput explosives-detection device (mostly high-throughput, noncertified, x-ray explosives-detection devices) with a lower P_d than a certified EDS; the first step in the DF approach is CAPS. For both DF and TPF systems, the second step is the screening

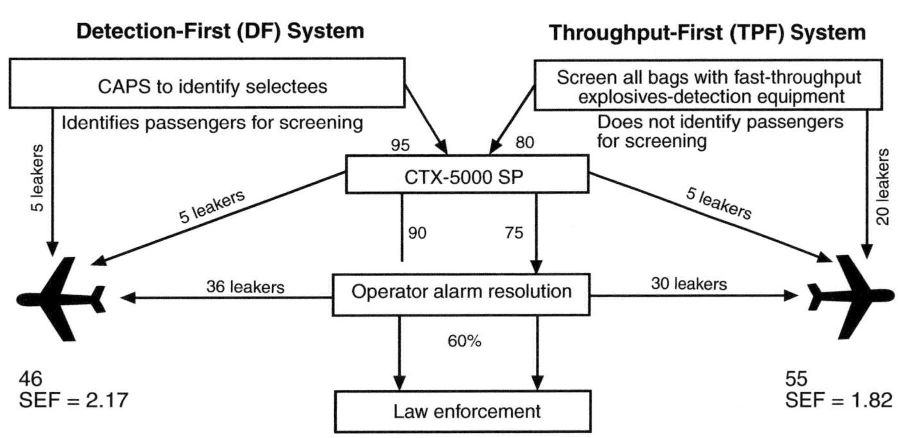

FIGURE 10-3 Hypothetical performance of detection-first and throughput-first aviation security systems for 100 MBTSs.

BOX 10-1
A Notional Example of the Impact of a Detection-First System on the Security Enhancement Factor

For the purposes of this example, a *baseline* TAAS is assumed to provide no security procedures or equipment at all ($P_d = 0$). Thus, if 100 MBTSs were inserted into the passenger flow, all 100 of them would pass through the baggage-handling system and be placed aboard aircraft.

Assume that the CAPS system can identify 90 percent of all passengers as very unlikely to be simulated terrorists (STs). Then assume that the STs insert 100 MBTSs into the passenger flow for a blind test of the *improved* TAAS. Ten percent of the total passengers will thus be declared selectees, and their bags will receive further scrutiny. Assume that as many as 5 percent (5) of the STs have been misidentified by CAPS as *nonthreat* passengers and that their bags will not be further inspected. If each ST is carrying one bag containing a simulated explosive, then five bags containing simulated explosives will not be further inspected. Of the bags sent by CAPS for further inspection by an EDS, 95 will contain simulated explosives.

Assume that the EDS fails to detect approximately 5 percent (5) of the MBTSs sent for inspection. Before the operator resolves a single alarm, a total of 10 percent (10) of ST MBTS bags would have been missed (5 missed by CAPS and 5 missed by the EDS). Therefore, approximately 90 percent (~90) of the threat bags (i.e., bags containing simulated explosives) would require alarm resolution by the EDS operator. Of the many bags that cause alarms, the operators must find 90 bags that contain simulated explosives. Assume that the operator (because he or she is rushed, and it is difficult to resolve one bomb out of thousands of false alarms) detects only 60 percent of the 90 bags that triggered EDS alarms that actually contain simulated explosives. This means that 36 bags containing simulated explosives are cleared. The security system would have missed 36 + 5 + 5 = 46 MBTSs, for a P_d of 54 percent. This is a significant improvement over the 100 missed bombs of the *baseline* TAAS. Therefore, SEF =100/46 = 2.17 or a twofold improvement.

by a certified EDS (e.g., CTX-5000 SP) of bags not cleared in step one. The alarm-resolution procedures are similar for both approaches; and passenger screening for weapons and the inspection of hand-carried baggage are also similar.

Detection-First Approach

Improvements in aviation security from the deployment of new security components (e.g., TEDDs or EDSs) are not directly related to their performance. All of the components of the TAAS must be considered together, either in a total-system blind test or by systems analysis to determine the SEF. The utilization rate of TAAS components, their reliability, the regulatory guidance provided by the FAA for their use, and human factors also figure into the SEF. A terrorist's perception of the effectiveness of the TAAS must also be considered because, ultimately, the terrorist's perception of the TAAS, rather than the actual performance, determines the level of deterrence. See Box 10-1 for a notional example of how the SEF can be used to estimate the performance of deployed equipment.

Hardened Unit-Loading Devices

A HULD for wide-body jets (e.g., the Boeing 747) is currently being operationally tested by three airlines. One HULD is probably large enough to hold all of the selectee bags for a given flight (note that it is unlikely that more than one selectee bag would contain a bomb, or bomb test set, on a given flight). A bomb at the minimum threat weight would be contained in the HULD, which has a high probability of saving the aircraft; thus, the HULD would improve the SEF. In the notional example in Box 10-1, 46 out of 100 passenger bags that contain MBTSs would be missed by the baseline DF security system. Five of the 46 bags were missed by CAPS (i.e., five simulated terrorists were not identified as selectees) and, therefore, the use of HULDs to hold selectee bags would not address these five bags. However, all 41 selectee bags not detected or improperly cleared by explosives-detection equipment would be contained in a HULD (although probably not all in the same HULD). The five bags containing MBTS devices that were not selected would not be detected or put into the HULD. Thus, the SEF in Box 10-1 would be 100/5 = 20.

The proper combination of technologies in the right order can dramatically improve the SEF. EDSs can more easily detect larger bombs (i.e., the larger the bomb the higher the P_d), whereas the HULD becomes less effective as the size of the bomb increases. A combination of EDSs and HULDs can, therefore, reduce the overall risk of aircraft loss. With an EDS in place, a terrorist would be forced to use smaller bombs; without an EDS, the terrorist might use a larger bomb than the HULD could contain. Thus, EDSs and HULDs are mutually dependent for improving aviation security.

Substantial improvement of the TAAS might involve trade-offs between some parameters of EDS performance under operational conditions if the TAAS includes HULDs. The optimum security system might trade off P_d, P_{fa}, and throughput rate with the capability of the HULD to contain explosives of a particular size.

Trace Explosives-Detection Devices

Most passengers in the western hemisphere fly on narrow-body aircraft (e.g., Boeing 717, 727, 737, 757). At present, no HULDs are available for use on narrow-body aircraft. Therefore, improving the SEF for flights on narrow-body aircraft would require a substantial improvement in the ability of EDS operators to detect explosives in alarmed bags. The FAA has deployed TEDDs with some EDSs to assist operators in resolving alarms. The TEDD can be used to sample electronic devices (e.g., laptop computers, radios) that are hard to resolve when the bag is opened and manually inspected. If it is assumed that 25 percent of the bags mistakenly cleared by an operator contain bombs concealed in electronic devices, a TEDD should reduce the number of bags mistakenly cleared. In the example in Box 10-1, 36 of the 100 bags that contain a test bomb (simulated explosive) are mistakenly cleared by the operator. Therefore, if 25 percent of the 36 bags the operator mistakenly cleared were opened and tested with a TEDD, nine of them would be sampled with a TEDD. In this example, the electronic device is assumed to be contaminated with a detectable level of explosive material, and the P_d of the TEDD is about 80 percent. Thus, eight of the nine simulated explosives would be identified. Thus, using a TEDD with each EDS would improve the SEF (SEF = 100/(5 [missed by CAPS] + 5 [missed by EDS] + 36 [mistakenly cleared by EDS operator] – 8 [detected by TEDDs]) = 100/38 = 2.63) by about 20 percent. If the protocol were altered so that every bag alarmed by the EDS were screened with a TEDD, each of the 90 bags containing test bombs that would normally be alarmed by the EDS (in the example in Box 10-1) would be subsequently subjected to a TEDD. For this example, it is assumed that the operator still detects 60 percent of the MBTSs alarmed by the EDS (i.e., 54 would be detected by the operator, and 36 would be missed). The TEDD would detect 80 percent of the 36 missed by the EDS operator (i.e., 28 would be detected and 8 would be missed). Thus the SEF = 100/(5 [missed by CAPS] + 5 [missed by EDS] + 8 [missed by TEDD and EDS operator]) = 5.6.

Operator Training

In the notional example in Box 10-1, 46 out of 100 passenger bags that contain test bombs would be missed by the DF security system. Five of the 46 bags that were missed were missed by CAPS (i.e., five simulated terrorists were not identified as selectees), an additional five were missed by the EDS, and 36 were mistakenly cleared by the EDS operator. If the FAA required operators to achieve an 80 percent detection level for MBTSs alarmed by an EDS, only 28 (i.e., 20 percent of 90 bags containing MBTS alarmed by EDS [18] + 5 missed by CAPS + 5 missed by EDS) MBTSs in the example in Box 10-1 would be missed. The SEF would be 100/28 = 3.6.

Throughput-First Approach

If potential terrorists cannot be identified by passenger screening,[1] it may be necessary to screen all bags. TPF systems rapidly screen all bags. Typically, a noncertified explosives-detection device is used for the initial screening; the P_d is as high as possible while maintaining a high throughput rate. The P_d of these devices is lower than the P_d of certified EDSs, but their throughput rate is much higher. If a bag cannot be cleared, it is sent to a more sensitive system, such as the FAA-certified CTX-5000 SP, where an operator can take more time to resolve the alarm. An example of the impact of a TPF system on SEF is shown in Box 10-2. The effects on the TAAS of DF and TPF systems are shown in Table 10-1 (based on the same assumptions used in Boxes 10-1 and 10-2).

Comparison of Detection-First and Throughput-First Approaches

The analysis presented in this chapter is based on the assumption that the rate of initial detection using CAPS in a DF system will be higher than with a bulk explosives-detection device in a TPF system. Thus, the key parameter in comparing DF and TPF approaches is how well the *first level* (i.e., CAPS vs. a high-throughput explosives-detection device) detects a threat. As long as CAPS is more effective, the DF approach will have a higher SEF than the TPF approach (see Figure 10-1). Note that no quantitative evidence is available to indicate whether CAPS is more or less effective than bulk explosives-detection equipment.

CAPS is less costly than screening all bags and reduces the complexity of the TAAS, but the capability of screening passengers and not missing potential terrorists has not been tested. The U.S. air transport system has a very large domestic component and many ways to identify nonthreatening passengers (e.g., airline frequent-flyer programs). The impact on the SEF of CAPS in combination with various other security measures is shown in Figure 10-4.

CONCLUSIONS AND RECOMMENDATIONS

Preliminary analyses suggest that the deployment and utilization of new security equipment could substantially

[1] The airlines with the highest potential terrorist threat believe that interviewing is the best way to identify passengers with baggage that may contain a bomb.

BOX 10-2
A Notional Example of the Impact of a Throughput-First System on Security Enhancement Factor

For the purpose of this example, it is assumed that a *baseline* TAAS provides no security procedures or equipment at all ($P_d = 0$). Thus, if 100 MBTSs were inserted into the passenger flow, all 100 of them would pass through the baggage-handling system and be placed onboard aircraft.

For a blind test of the TPF-based TAAS, simulated terrorists (STs) insert 100 MBTSs into the passenger flow. Assume that the P_d of the explosives-detection equipment used for initial screening of baggage in a TPF approach is 80 percent and that the probability of missing an explosive is 20 percent. Thus, initially the TPF approach will miss 20 percent (20) of the MBTS bags. Assume that the certified EDS misses 5 percent of the 80 bags that were sent on for further scrutiny (4 bags) and that EDS operators miss another 40 percent of the 76 bags alarmed by the EDS (31 bags). The SEF would then be 100/(20 + 4 + 31) = 100/55 = 1.82. Therefore, the TPF TAAS is an improvement over the *baseline* TAAS.

improve aviation security as measured by the SEF. Security would be enhanced more where the threat is highest. The performance of the TAAS can only be assessed with systems analysis techniques that account for the contribution and sensitivities of each component. The certification and optimization of a subsystem (e.g., EDS) do not necessarily optimize the TAAS. The overall system must be tested frequently in airport operations by blind testing procedures to provide data for the system analyses and to train operators.

Analysis of the Total Architecture for Aviation Security

Most data on deployed aviation security systems are incomplete. The TAAS performance used for the preliminary systems analysis described in this chapter was based on very limited quantitative data and some anecdotal information. However, the examples indicate the types of data necessary for a more thorough TAAS-level analysis of aviation security. The panel concluded that data should be collected over a range of conditions to evaluate the impact of component performance (e.g., EDSs, TEDDs) on TAAS performance. Data collection should be continuous, professionally staffed, and well funded.

Recommendation. The FAA should design and implement a mechanism for collecting data to ensure that sufficient data are available for substantive analyses of the total architecture for aviation security.

Recommendation. The FAA should collect comprehensive total architecture for aviation security performance data based on operational blind testing as a basis for measuring improvements in total architecture for aviation security performance levels under airport operational conditions.

The discussion of TPF and DF approaches in this chapter was based on the assumption that CAPS is a more effective way to identify threat passengers (i.e., passengers with explosives concealed in their baggage) than a high-throughput explosives-detection device was in finding explosives concealed in baggage. However, a thorough comparison of the TPF and DF approaches has not been completed. The panel concluded that an analysis that includes measures of the effectiveness of CAPS and high-throughput explosives-detection equipment should be done to determine the relative impact of the two approaches on overall aviation security.

TABLE 10-1 Potential Improvements in the SEF for Detection-First and Throughput-First Aviation Security Systems

TAAS Configuration	Throughput-First System	Detection-First System
Notional system (depicted in Figure 10-2)	1.82	2.17
Notional system + TEDD	2.04	2.63
Improve operator training in identifying MBTSs	2.5	3.6
Operator training + TEDD	2.7	4.2
HULDs added to the deployed system	5.0	20

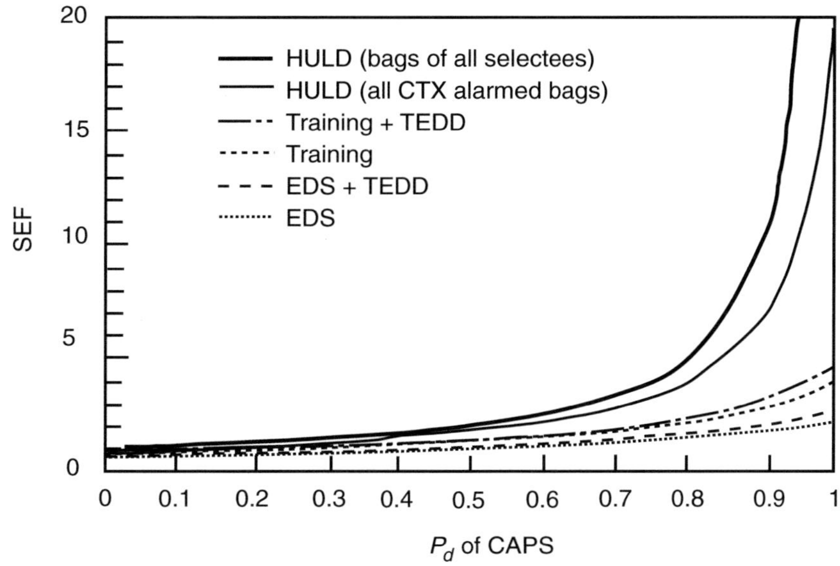

FIGURE 10-4 Notional values of the SEF as a function of the efficiency of CAPS in combination with other security measures.

Recommendation. The FAA should conduct a thorough analysis of the effectiveness of throughput-first and detection-first approaches. The security enhancement factor (or a similar measure) should be used as a basis for comparison.

Recommendation. To increase the deterrent effect of the overall security system, analysis, testing, and training should be visible to the public. The detection of simulated terrorists threats should be publicly acknowledged but with only enough detail to create uncertainty about the performance of the overall security system in the terrorist's mind.

TAAS Components

Although the FAA has not defined how a HULD should be used to increase security, the simple analysis in this chapter suggests that using one HULD on each wide-body aircraft would increase the SEF almost tenfold. The panel concluded that the greatest benefit would be derived if the bags of all CAPS selectees were placed in a HULD, even if their bags were subsequently cleared by other security procedures.

Recommendation. When hardened unit-loading devices (HULDs) become more widely available and can be integrated into air carrier systems without unreasonably disrupting operations, the FAA should require that all selectee bags for wide-body aircraft be placed in HULDs. A similar requirement should be established for narrow-body aircraft when HULDs for them become available.

The operational performance data available to the panel for TEDDs were anecdotal and, therefore, not sufficient to establish the optimum use of TEDDs in the context of the TAAS. However, even anecdotal and superficial data suggest that TEDDs do improve the SEF.

Recommendation. The FAA should develop procedures and test capabilities to assess the operational performance of trace explosives-detection devices in an airport environment.

Security Enhancement Factor

The SEF described in this report is based on minimizing false negatives for the TAAS as a whole. The SEF (or a similar measure) can also be used as a tool to assess the effectiveness of various components of the TAAS.

Recommendation. The FAA should develop a systems analysis to evaluate the total architecture for aviation security in terms of a security enhancement factor (SEF) or a similar measure. As a first step, the FAA should use the data from blind testing. The SEF estimated by these results could be used to assess the benefits of newly deployed systems.

Recommendation. Decisions to deploy new equipment and implement new procedures should be based on a full analysis of the total architecture of aviation security, including the security enhancement factor or a similar measure.

11

Response to Congress

In keeping with the Statement of Task for this study, the panel reviewed the FAA's deployment of technologies to improve aviation security, including explosives-detection equipment and HULDs. Given the complexity of the terrorist threat and the short response time mandated by Congress, the deployment has been successful. Considering that hardware had to be deployed immediately, the FAA did not have sufficient time to follow a systems analysis and design protocol to achieve an optimum systems architecture. Therefore, at this stage, the lack of comprehensive data for evaluating the effectiveness of the deployed equipment is not surprising. However, the FAA must now address the issues of collecting performance data and systems integration.

Explosives-detection equipment and HULDs are part of a TAAS (total architecture for aviation security), and the panel strongly believes that they should be evaluated in that context in this and in future NRC studies. Although both the FAA and its contractors have adopted some aspects of this systems approach, they have not characterized the optimal deployment scenario. With the systems approach, improvements in security can be characterized by an SEF.

Based on available data and current understanding of the various elements of the system, the panel was able to represent a notional analytical summary of security enhancement as a function of current and projected system performance and concluded that substantial technical improvements in security have already been made. As more data are collected and analyzed, Figure 10-1 can be refined and improved.

The SEF also includes the resolution of detected alarms, which requires that operators make decisions based on the full capability of the hardware. For this reason, operator training could have a significant effect on the SEF. Based on the statistically significant probability that an alarmed bag containing an explosive device might be "cleared" by an operator, the panel determined that an order of magnitude increase in the SEF could result if all alarmed bags cleared by an operator (during the alarm-resolution process) were placed in a HULD (see Chapter 10).

The following overarching recommendations address the four issues raised by Congress.

1. Assess the weapons and explosives-detection technologies available at the time of the study that are capable of being effectively deployed in commercial aviation.

This study focused on explosives-detection technologies. While it is conceivable that some of these technologies could also be used for weapons detection, this topic was not addressed in this report.

BULK EXPLOSIVES-DETECTION EQUIPMENT

The vast majority of bulk explosives-detection equipment deployed is the FAA-certified InVision CTX-series EDS. Most of the performance data on this equipment was generated during laboratory testing—largely certification testing—at the FAA Technical Center. Certification tests, however, only reflect the ability of the equipment to detect a bag that contains an explosive, and the detection rates are based on bag-alarm rates. That is, an explosive is considered to be detected if the alarm is set off for the bag containing the explosive, even if the alarm is triggered by a nonexplosive object in the bag. Certification testing does not measure alarm resolution and does not include testing in the operational environment of an airport, making it difficult to evaluate the operational performance of explosives-detection technologies considered for deployment. In the panel's opinion, some of the unanticipated problems encountered with the CTX-5000 SP in the field can be reasonably related to the limitations of certification testing. Under current certification guidelines, equipment certified in the future may encounter similar problems.

Recommendation. During certification testing, the FAA should, whenever possible, measure both true detection rates (i.e., correctly identifying where the explosive is when an alarm occurs) and false-detection rates (i.e., an alarm triggered by something other than the explosive in a bag that contains an explosive). The FAA should also include the ability of explosives-detection equipment to assist operators in resolving alarms (including in an airport) as part of certification testing. Alarm resolution should be included in the measurement of throughput rate, detection rate, and false-alarm rate.

TRACE EXPLOSIVES-DETECTION DEVICES

TEDDs are widely used in airports, but no comprehensive methodology has been developed to evaluate their effectiveness, such as standard test articles or instrument and operator requirements. Because no standard test articles for TEDDs have been demonstrated—and because of the resultant inability to separate instrument and operator performance—it is not possible to measure the performance of TEDDs.

Recommendation. The FAA should develop and implement a program to evaluate the effectiveness of deployed trace explosives-detection devices. This evaluation should include measurements of instrument and operator performance, including measurements in the deployed (i.e., airport) environment.

COMPUTER-ASSISTED PASSENGER SCREENING AND POSITIVE PASSENGER-BAG MATCHING

CAPS appears to be an effective method of screening passengers to identify selectees for further security measures, such as bag matching or bag screening. The panel believes that CAPS is an effective method of focusing resources (personnel and equipment) on high-risk passengers that does not impede the operation of the air carriers. No quantitative measure of the effectiveness of CAPS has been demonstrated, however. The panel believes that testing the performance of CAPS will be necessary for its long-term viability.

The 1997 White House Commission on Aviation Safety and Security recommended that the FAA begin implementation of full passenger-bag matching. The FAA interpreted this recommendation to mean that passengers who were either randomly selected or who were identified through a profiling system (e.g., CAPS) should be subject to bag matching. The panel anticipates that PPBM combined with CAPS will be an effective tool for improving aviation security. However, under the current deployment scenario, when a selectee changes planes (or simply deplanes during a stopover) at a connecting airport, PPBM is not repeated.

Recommendation. Computer-assisted passenger screening (CAPS) should continue to be used as a means of identifying selectee passengers whose bags will be subject to positive passenger-bag matching (PPBM), screening by explosives-detection equipment, or both. PPBM combined with CAPS should be part of the five-year plan recommended below. Passengers designated as selectees at the origination of their flights should remain selectees on all connecting legs of their flights. Within six months, the FAA should develop and implement a method of testing the effectiveness of CAPS.

2. **Determine how the technologies referred to in paragraph (1) could be used more effectively to promote and improve security at airport and aviation facilities and other secured areas.**

PROGRESS IN THE DEPLOYMENT OF AVIATION SECURITY EQUIPMENT

As directed by Congress in Section 305 of the FAA Reauthorization Act of 1996 (PL 104-264) and recommended by the White House Commission on Aviation Safety and Security, the FAA has initiated the deployment of bulk and trace explosives-detection equipment, as well as the implementation of CAPS and PPBM. These measures have significantly improved aviation security. Nevertheless, the FAA/SEIPT is behind schedule in its deployment of aviation security equipment. Congress provided $144.2 million in the Omnibus Consolidated Appropriations Act of 1997 for the purchase of commercially available security screening equipment, and the FAA/SEIPT planned to deploy 54 certified explosives-detection systems and 489 trace-detection devices by December 1997. The FAA also planned to fully implement CAPS by December 1997. Once it became apparent that these goals could not be met, the FAA set a new goal of deploying 54 certified EDSs, 22 noncertified bulk explosives-detection devices, and 489 trace-detection devices by December 31, 1998. The FAA also planned to implement CAPS fully by December 31, 1998. As of January 1, 1999, more than 70 certified EDSs, six noncertified bulk explosives-detection devices, and 366 TEDDs had been installed in airports. In addition, 10 HULDs have been deployed to three airlines for operational testing. Thus, the FAA has made significant progress towards achieving its updated goals. Nevertheless, there is not a sufficient amount of deployed equipment to screen every passenger bag or, in some cases, even every selectee bag without interrupting passenger flow.

The panel concluded that the FAA/SEIPT, the airlines, airports, and associated contractors have gained significant experience from the initial deployment of security equipment and procedures, and the current implementation of security equipment does not appear to have interfered unreasonably with airline operations. Most importantly, in the collective opinion of the panel, the deployment of security equipment has improved aviation security. The panel believes that continued emphasis on, funding of, and deployment of security equipment will further enhance aviation security. Future deployments should be more efficient if they are based on the experience from the initial deployment.

Recommendation. The U.S. Congress should continue to fund and mandate the deployment of commercially available explosives-detection equipment through the FAA/SEIPT. Continued deployments will increase the coverage of domestic airports and eventually provide state-of-the-art security equipment systemwide. Further deployments can improve aviation security in the short term and provide the infrastructure for mitigating potential threats in the long term.

OPERATOR PERFORMANCE

Human operators are integral to the performance of all deployed explosives-detection equipment. Because fully automated explosives-detection equipment will not be developed in the foreseeable future, particularly with respect to alarm resolution, human operators will continue to be immensely important to realizing the full potential of deployed security hardware. The TAAS analysis presented in this report quantifies the impact of the operator on the SEF. Certification testing of explosives-detection equipment, however, does not include testing of human operators. Current testing only defines the operational capability (or performance) of the equipment.

It has been demonstrated that the introduction of an operator (who is needed to resolve alarms) decreases the effectiveness of deployed explosives-detection systems. FAA test results indicate that, even if explosives-detection equipment correctly identifies a potential threat, the operator can make an incorrect decision and "clear" a bag with a bomb in it. The P_d of the equipment/operator combination is, therefore, lower than the P_d of the equipment alone.

Recommendation. The FAA should institute a program to qualify security-equipment operators to ensure that the human operator/explosives-detection system (EDS) combination meets the performance requirements of a certified EDS. This program should include the definition of operator performance standards and a means of monitoring operator performance. The FAA should implement this program within six months of receipt of this report.

MEASURING OPERATIONAL PERFORMANCE

Because of the paucity of operational data for deployed explosives-detection equipment, the panel found it impracticable to characterize the deployment status of security equipment and processes quantitatively. The data are insufficient both for the equipment and for operator performance, and no quantitative measures of the effectiveness of the total security system (e.g., TAAS) were provided to the panel. The majority of data focused on subsystems, such as bulk explosives-detection systems. A thorough assessment of equipment and system performance requires well defined performance metrics and the collection of data. The panel concluded that the FAA has not defined adequate performance metrics for security subsystems (e.g., TEDDs) or for the TAAS.

Recommendation. The FAA should make a concerted effort to define operational performance metrics for security subsystems and for the total architecture for aviation security (TAAS). The FAA should also create an action team in the next six months to systematically collect operational data, which should be used to optimize the TAAS, as well as to identify and correct substandard performance of equipment and operators. The data collected would also provide insights into the deployment and use of equipment in the future.

MEASURING SECURITY ENHANCEMENT

Besides the lack of operational data and the apparent lack of performance metrics for the total system, no overall measure of security enhancement has been defined. Because the primary performance measure for the TAAS is, of course, protection against explosive threats, the critical factor for assessing the performance of the TAAS is the leakage of false negatives (i.e., bags that contain bombs). The panel defined improvements in performance, or the SEF, as the number of simulated bombs that defeat a baseline security system divided by the number of simulated bombs that defeat the newly deployed system. The panel believes that the SEF is essential to assessing the impact of security equipment and procedures.

Recommendation. The FAA should formulate a security enhancement factor (SEF) for the integrated total architecture for aviation security systems. The SEF should be calculated from data collected during operational testing. Nonclassified SEF measures should be published and used as a project-control and management-control tool. The SEF would provide the FAA with a quantitative measure of the impact of security equipment and procedures.

FIVE-YEAR DEPLOYMENT PLAN

TAAS-based decisions involve airport and airline-specific management and costs. Airline and airport buy-in are, therefore, necessary for effective deployment. The FAA has not demonstrated to the panel a long-range (five-year) TAAS deployment plan developed jointly and agreed to by the stakeholders. Thus, to the knowledge of this panel, the airlines have not agreed to a long-term deployment plan that addresses all of the relevant issues, such as operator training, the optimal location of detection equipment, and the deployment of HULDs. The panel believes that the development of a long-term deployment plan is critical for continued improvements in aviation security.

Recommendation. Within one year, in cooperation with the other stakeholders, the FAA should develop a five-year joint-deployment plan that includes cost, stakeholder responsibilities, quality measures, and other important factors. This plan should be a living document that is formally updated annually. Buy-in from all stakeholders will be necessary for the plan to be effective.

3. Assess the cost and advisability of requiring hardened cargo containers to enhance aviation security and reduce the required sensitivity of bomb-detection equipment.

Two HULDs (both LD-3 size) have passed the FAA blast and shockholing tests and conform to the NAS-3610-2K2C airworthiness criterion. However, U.S. airlines have resisted the deployment of HULDs on the basis of operational concerns. The effects of operational wear and tear on the blast-containment characteristics of HULDs must be tested operationally.

The HULDs that passed the FAA blast test are the size of an LD-3, a container size used on wide-body aircraft. No HULD concepts for narrow-body aircraft have been developed. More than 70 percent of bombing attempts have been against narrow-body aircraft, and 75 percent of the aircraft in service (as of 1994) are narrow-body aircraft. Therefore, before the FAA recommends, mandates, or regulates the use of hardened containers for airline operations, the feasibility of HULDs designed for narrow-body aircraft or, perhaps, other protection concepts will have to be further investigated.

Recommendation. The FAA should continue to support research and development on hardened unit-loading devices (HULDs), including ongoing operational testing. If the FAA recommends, mandates, or regulates the use of HULDs, explosion-containment strategies for narrow-body aircraft, including the development of narrow-body HULDs and cargo-hold hardening concepts, should be investigated. However, the FAA should not deploy HULDs unless they are part of the TAAS joint five-year deployment plan.

4. On the basis of the assessments and determinations made under paragraphs (1), (2), and (3), identify the most promising technologies for improving the efficiency and cost effectiveness of weapons and explosives detection.

The data were not sufficient for a comprehensive assessment of available technologies for improving aviation security. Therefore, at this time, the panel is not able to identify or recommend the most promising technologies for improving the efficiency and cost effectiveness of weapons and explosives detection. If the recommendations in this report are followed, these data will become available for subsequent assessments.

References

Aven, T. 1992. Reliability and Risk Analysis. New York: Elsevier.

British Air Accident Investigation Branch. 1990. Report on the Accident to Boeing 747-121, N739PA at Lockerbie, Dumfriesshire, Scotland on 21 December 1988. Farnborough, U.K.: British Air Accident Investigation Branch.

Dickey, R. 1998. Panel on Assessment of Deployed Technology, SEIPT Evaluation Team. Briefing by R. Dickey, Evaluation Product Team Leader on the SEIPT, to the Panel on Assessment of Technologies Deployed to Improve Aviation Security, National Research Council, Washington, D.C., January 29, 1998. (Section 191 Restricted)

Dickey, R., and D. Fuqua. 1998. Evaluation of the FAA's Security Equipment Integration Program. Presented by R. Dickey at the International Test and Evaluation Association (ITEA) Symposium, September 21, 1998, Albuquerque, New Mexico.

DOT (U.S. Department of Transportation). 1997. Test and Evaluation Plan for Explosive Trace Detection Equipment Study. DOT/FAA/AR-97. Washington, D.C.: U.S. Department of Transportation.

DOT. 1998. Aviation Security: Federal Aviation Administration. Washington, D.C.: U.S. Department of Transportation, Office of the Inspector General. Available on line at: *http://www.dot.gov/oig/audits/av1998134.html*

Drury, C.G. 1989. The Human Operator as Inspector: Aided and Unaided. Pp. A65–A71 in Proceedings of Human Factors Issues in Aircraft Maintenance and Inspection. DOT/FAA/AAM-89/9. Washington, D.C: U.S.: Federal Aviation Administration, Office of Aviation Medicine.

FAA (Federal Aviation Administration). 1992. Proposed Criteria for Certification of Explosives Detection Systems as Defined under 14 CFR 108.20. Washington, D.C.: Federal Aviation Administration.

FAA. 1995. Technical Center Baggage Inspection System: Airport Operational Demonstration Project. Washington, D.C.: Federal Aviation Administration.

FAA. 1997a. CTX-5000 and X-ray Blind Testing at San Francisco International Airport. Washington, D.C.: Federal Aviation Administration.

FAA. 1997b. Interim Report on the CTX-5000 SP Airport Demonstration. Washington, D.C.: Federal Aviation Administration.

FAA. 1997c. Final Report on CTX-5000 SP Operational Testing at Atlanta. Washington, D.C.: Federal Aviation Administration.

FAA. 1997d. DOT/FAA Solicitation DTFA03-97-R-00008, Phase II Certified Blast Mitigating Airline Baggage Containers. 29 January 1997. Washington, D.C.: Federal Aviation Administration.

FAA. 1998. Engineering Design Approval for ECOS Hardened Luggage Container. Letter from FAA Transport Airplane Directorate, Certification Office, to Galaxy Corporation. July 29, 1998.

Fainberg, A. 1998. Strategic planning for aviation security. Briefing by A. Fainberg, director, FAA Office of Civil Aviation Security Policy and Planning, to the Panel on Assessment of Technologies Deployed to Improve Aviation Security. National Research Council, Washington, D.C., July 21, 1998. (Section 191 Restricted)

FBI (Federal Bureau of Investigation). 1995. Terrorism in the United States in 1995. FBI Available on line at: *http://www.fbi.gov/publish/terror/terrorin.htm*

Fox, F. 1998. Introduction to trace explosives detection instruments. Briefing by F. Fox, FAA research chemist, to the Panel on Assessment on Technologies Deployed to Improve Aviation Security. National Research Council, Washington, D.C., July 22, 1998.

Fuqua, D. 1998. Assessment concept for explosive-detection equipment and processes. Briefing by D. Fuqua, Scientific Research Corporation, to the Panel on Assessment on Technologies Deployed to Improve Aviation Security. National Research Council, Washington, D.C., May 1, 1998. (Section 191 Restricted)

Fuqua, D., and T.G. O'Brien. 1998. SEIPT Operational Assessment Report (March 20, 1998). Washington, D.C.: Federal Aviation Administration.

GAO (General Accounting Office). 1998. Aviation Security: Implementation of Recommendations Is Under Way, But Completion Will Take Several Years. GAO/RCED-98-102. Washington, D.C.: General Accounting Office. Available on line at: *http://www.gao.gov/AIndexFY98/abstracts/rc98102.htm*

Greshem, G.L., J.P. Davies, L.D. Godorich, L.G. Blackwood, L.Y. Benjamin, D. Thimsem, S.E. Yoo, and S.F. Hallowell. 1994. Development of Particle Standards for Testing Detection Systems: Mass of RDX and Particle Size of Distribution of Composition 4 Residues. International Society for Optical Engineers Meeting Proceedings 2276: 34–44.

Hacker, K. 1998a. Hardened container program. Briefing by K. Hacker, FAA, to the Panel on Assessment of Technologies Deployed to Improve Aviation Security. National Research Council, Washington D.C., January 30, 1998. (Section 191 Restricted)

Hacker, K. 1998b. Hardened container program (update). Briefing by K. Hacker, FAA Aircraft Hardening Program Manager, to the Panel on Assessment of Technologies Deployed to Improve Aviation Security. National Research Council, Washington, D.C., July 21, 1998. (Section 191 Restricted)

Hammar, P. 1998. Deployment analysis database. Briefing by P. Hammar, director, security and system analysis, DSCI, to the Panel on Assessment of Technologies Deployed to Improve Aviation Security. National Research Council, Washington, D.C., July 21, 1998.

Harris, D.H., and F.B. Chaney. 1969. Human Factors in Quality Assurance. New York: John Wiley and Sons.

Henly, E.J., and H. Kumamoto. 1992. Probabilistic Risk Assessment. New York: IEEE Press.

Hou, T.-H., L. Lin, and C.G. Drury. 1994. Development and Evaluation of

Hybrid Inspection Systems. Pp. 684–687 in Advances in Agile Manufacturing, P.T. Kidd and W. Karwowski (eds.). Washington, D.C.: IOS Press.

IATA (International Air Transport Association). 1995. Hardened Containers: Cost and Operational Impact Estimates for the International Airline Industry. Draft Position Paper. Washington, D.C.: Unit Load Devices Technical Board.

INEL (Idaho National Energy Laboratory). 1997. Test Plans for the Independent Verification and Validation of Trace Explosive Detection Systems for the FAA. INEL/EXT-97-00377. Washington, D.C.: Federal Aviation Administration.

INEL 1998. Results of the Independent Verification and Validation of Trace Explosive Detection Systems for the FAA. INEEL/EXT-98-Draft. Washington, D.C.: Federal Aviation Administration.

Karasek, R., and T. Theorel. 1990. Healthy Work. New York: Basic Books.

LaMonica, D., J. McNeill, L.R. Powell, B. Scavetta, J. Janowitz, D. Yeager, and A. Lupinetti. 1995. Aviation Security Technology Integration Plan. Atlantic City, N.J.: FAA Technical Center (by contract to Galaxy Scientific Corporation).

McDonnell Douglas. 1997. Development of Aircraft Structural Response Methodology. St. Louis, Mo.: McDonnell Douglas Corporation.

Morrocc, J.D. 1997. Hardening concepts tested to counter terrorist blasts. Aviation Week and Space Technology 146 (23): 44–45.

National Institute for Aerospace Studies and Services. 1996. Aircraft Response to Internal Explosive Detonations, Final Report. October 1996. Washington, D.C.: National Institute for Aerospace Studies and Services.

Neiderman, E.C., and J.L. Fobes. 1997. A Cognitive Model of X-Ray Security Screening: Selection Tests to Identify Applicants Possessing Core Aptitude. Final Report. DOT/FAA/AR-97/63. Springfield, Va.: National Technical Information Service.

NRC (National Research Council). 1996a. First Interim Report of the Committee on Commercial Aviation Security. National Materials Advisory Board. Washington, D.C.: National Academy Press.

NRC. 1996b. Airline Passenger Security Screening. National Materials Advisory Board. Washington, D.C.: National Academy Press. Available on line at: *http://www.nap.edu/readingroom/enter2.cgi?0309054397.html*

NRC (National Research Council). 1997. Second Interim Report of the Committee on Commercial Aviation Security. National Materials Advisory Board. Washington, D.C.: National Academy Press. Available on line at: *http://www4.nas.edu/cets/nmab/*

Padgett, J. 1998. Computer-aided passenger screening. Briefing to the Panel on Assessment of Technologies Deployed to Improve Aviation Security. National Research Council, Washington, D.C., July 22, 1998. (Section 191 Restricted)

Polillo, R. 1998. SEIPT strategy for integrating advanced technology into operation. Briefing by R. Polillo, leader, SEIPT, to the Panel on Assessment of Technologies Deployed to Improve Aviation Security. National Research Council, Washington, D.C., July 22, 1998. (Section 191 Restricted)

Reason, J. 1990. Human Error. Cambridge, U.K.: Cambridge University Press.

Rork, S. 1994. Personal communication from S. Rork, National Air Transport Association, to Bruce Butterworth, Office of Civil Aviation Security, Federal Aviation Administration, September 8, 1994.

Schwartz, E., P.A. Barr, and J.A. Gatto. 1995. Terrorist Bombings Against Commercial Aviation and Aircraft Hardening Research. Presented at AIAA, Aircraft Engineering, Technology, and Operations Congress, September 19–21, 1995, Los Angeles, California.

Sinclair, M.A., and C.G. Drury. 1979. On mathematical modeling in ergonomics. Applied Ergonomics 10(4): 225–234.

Walsh, W.J., and P.J. Yee. 1990. Embedded Training: Lessons Learned. Pp. 73 in Proceedings of the Topical Meeting on Advances in Human Factors Research on Man/Computer Interactions: Nuclear and Beyond. Nashville, Tenn.;. LaGrange Park, Ill.: American Nuclear Society.

White House Commission on Aviation Safety and Security. 1996. Initial Report to the President. September 9, 1996. Available on line at: *http://www.aviationcommission.dot.gov/*

White House Commission on Aviation Safety and Security. 1997. Final Report to the President. February 12, 1997. Available on line at: *http://www.aviationcommission.dot.gov/*

Zuckerman, M.B. 1996. Are order and liberty at odds? U.S. News and World Report 121(5): 64.

Biographical Sketches of Panel Members

Thomas S. Hartwick (chair) is retired manager of the satellite payload program and system design at TRW and an expert in the manufacture and deployment of complex defense and surveillance systems. Dr. Hartwick has several years of experience as manager of various organizations in the aerospace industry at Hughes Aircraft Company and Aerospace Corporation. His areas of research include sensors and imaging, especially optical communications, far-infrared lasers and their applications, and laser heterodyne radiometry. Since leaving the aerospace industry in 1995, Dr. Hartwick has served on a number of academic, government, and industrial boards as a technical manager. He also served on the National Research Council (NRC) Committee on Optical Science and Engineering.

Robert Berkebile is a consultant on air carrier operations with 41 years of experience. His area of expertise is in the coordination of cargo and passenger baggage. Before his retirement, Mr. Berkebile was manager of customer services automation for US Airways and director of station services for US-Africa Airways. He was a member of the NRC Panel on the Assessment of the Practicality of Pulsed Fast Neutron Transmission Spectroscopy for Aviation Security, under the Committee on Commercial Aviation Security.

Homer Boynton has extensive experience in security matters, including 25 years with the Federal Bureau of Investigation and 12 years with American Airlines as managing director of corporate security. He has chaired many advisory panels on airline security, including the Security Advisory Committee of the Air Transport Association and the Security Committee of the International Air Transport Association and was a member of the Federal Aviation Administration (FAA) Research, Engineering and Development Advisory Committee. Mr. Boynton is a member of the NRC Committee on Commercial Aviation Security.

Barry D. Crane of the Institute for Defense Analyses has a broad background in developing test and evaluation procedures for military hardware. He was responsible for developing drug-detection technology for the director of defense research and engineering, U.S. Department of Defense, until 1991. He is currently evaluating detection and monitoring capabilities of U.S. government surveillance assets used to find and follow air and maritime drug traffickers. He is also responsible for the evaluation of tactical aircraft testing for the director of operational test and evaluation, Office of the Secretary of Defense. Dr. Crane is a member of the NRC Committee on Commercial Aviation Security.

Colin Drury is professor of industrial engineering at the State University of New York at Buffalo and executive director of the Center for Industrial Effectiveness, where he has worked extensively on the integration of ergonomics/human factors into company operations to increase competitiveness and job growth. Since 1990, Dr. Drury has headed a team applying human factors to the inspection and maintenance of civil aircraft. He conducted a study for the Air Transport Association evaluating the FAA's modular bomb set and the use of this bomb set in training and testing security screeners. Dr. Drury is a fellow of the Human Factors and Ergonomics Society, the Institute of Industrial Engineers, and the Ergonomics Society. In 1981, he was awarded the Bartlett Medal by the Ergonomics Society and in 1992 the Paul Fitts Award by the Human Factors and Ergonomics Society. Dr. Drury has served on several NRC committees and panels.

Len Limmer retired in January 1998, closing a distinguished 26-year career with the Dallas/Fort Worth International Airport Board, a local government agency responsible for the administration, management, and operation of the world's second busiest airport. He has cross-functional experience in public safety, including security police, counterterrorism, explosives detection/disposal, crash rescue, and structural fire and hazardous-materials remediation. He is an officer of numerous Texas and national associations concerned with public safety in large metropolitan areas.

Harry E. Martz is leader of the nondestructive evaluation research and development thrust area at Lawrence Livermore National Laboratories (LLNL). For six years, he led the computed-tomography project at LLNL, applying computed-tomography and x-ray and proton radiography to material characterization and gamma-ray gauge techniques to treaty verification activities. His current projects include the use of nonintrusive x-ray and gamma-ray computed-tomography techniques as three-dimensional imaging tools to understand material properties and analyze radioactive waste forms. He has applied these techniques to the inspection of automobile and aircraft parts, reactor fuel tubes, high explosives, shaped charges, and the contents of waste drums. His research has focused on the design and construction of scanners and pre-processing, image reconstruction, and analysis algorithms. Dr. Martz chaired the NRC Panel on Configuration Management and Performance Verification of Explosives-Detection Systems and was a member of the Panel on Airport Passenger Screening and the Panel on the Assessment of the Practicality of Pulsed Fast Neutron Transmission Spectroscopy for Aviation Security. He is currently a member of the Committee on Commercial Aviation Security.

Joseph A. Navarro is president of JAN Associates. He has extensive experience in testing explosives-detection equipment in the laboratory and in the statistical analysis of test data. Dr. Navarro was instrumental in the development of the explosives-detection system testing protocol used for FAA certification tests and has been a technical advisor to several NRC committees and panels on aviation security.

Eric R. Schwartz is corporate vice president of quality at The Boeing Company and has participated in many studies of safety and survivability for commercial aircraft. Mr. Schwartz performed investigations and managed engineering analyses of structural failures due to terrorist bombings on Pan Am 747, UTA Airlines DC-10, and Itavia DC-9 aircraft and has developed research proposals for advanced aircraft structures. He is a recognized expert on terrorist acts against commercial aviation and has presented numerous technical papers to the FAA, the American Institute of Aeronautics and Astronautics, the U.S. Department of Defense, and international aviation authorities.

Elizabeth H. Slate is an associate professor in the School of Operations Research and Industrial Engineering, Cornell University. She has conducted extensive research into statistical analysis methodologies related to industrial engineering issues. She was awared a recent grant from the Semiconductor Research Corporation/National Science Foundation for a study of methods for modeling stochastic processes in semiconductor manufacturing. Dr. Slate is active in the American Statistical Association and many other societies and is the chair elect of the American Statistical Association Subsection on Statistical Computing.

Michael Story has been involved in the research, design, and commercialization of mass spectrometers for 30 years. He is a cofounder of the Finnigan Corporation and is currently technology assistant to the president of ThermoQuest Corporation. He was a member of the NRC Committee on Commercial Aviation Security (1988–1993) and chaired the Panel on Test Protocol and Performance Criteria. Mr. Story also served on the NRC Panel on Configuration Management and Performance Verification of Explosives-Detection Systems.

Technical Report Documentation Page

1. Report No.	2. Government Accession No.	3. Recipient's Catalog No.
DOT/FAA/AR-99/79		

4. Title and Subtitle	5. Report Date
Assessment of Technologies Deployed to Improve Aviation Security, First Report	November 1999
	6. Performing Organization Code National Research Council
7. Author(s) Committee on Commercial Aviation Security Panel on Assessment of Technologies Deployed to Improve Aviation Security	8. Performing Organization Report No. NMAB-482-5
9. Performing Organization Name and Address National Materials Advisory Board National Research Council 2101 Constitution Avenue, NW Washington, DC 20418	10. Work Unit No. (TRAIS) 11. Contract or Grant No. DTFA03-94-C-00068
12. Sponsoring Agency Name and Address Federal Aviation Administration Technical Center Aviation Security Research and Development Service Atlantic City International Airport, NJ 08405	13. Type of Report and Period Covered Final Report 14. Sponsoring Agency Code AAR-500

15. Supplementary Notes

The FAA's Contracting Officer's Technical Representatives (COTR) are Alan Novakoff and Paul Jankowski

16. Abstract

This study was conducted under the auspices of the National Materials Advisory Board's Committee on Commercial Aviation Security to assess the operational performance of explosives detection systems and hardened cargo containers in airports and compare that performance to their performance in laboratory testing, with a focus on ways to deploy these systems more effectively to improve aviation security.

17. Key Words	18. Distribution Statement
explosives detection systems, aviation security, explosives detection technology, security procedures	

19. Security Classif. (of this report)	20. Security Classif. (of this page)	21. No. of Pages	22. Price
Unclassified	Unclassified	68	

Form DOT F1700.7 (8ñ72) Reproduction of completed page authorized

Technical Report Documentation Page

1. Report No. DOT/FAA/AR-99/79	2. Government Accession No.	3. Recipient's Catalog No.
4. Title and Subtitle Assessment of Technologies Deployed to Improve Aviation Security, First Report		5. Report Date November 1999 6. Performing Organization Code National Research Council
7. Author(s) Committee on Commercial Aviation Security Panel on Assessment of Technologies Deployed to Improve Aviation Security		8. Performing Organization Report No. NMAB-482-5
9. Performing Organization Name and Address National Materials Advisory Board National Research Council 2101 Constitution Avenue, NW Washington, DC 20418		10. Work Unit No. (TRAIS) 11. Contract or Grant No. DTFA03-94-C-00068
12. Sponsoring Agency Name and Address Federal Aviation Administration Technical Center Aviation Security Research and Development Service Atlantic City International Airport, NJ 08405		13. Type of Report and Period Covered Final Report 14. Sponsoring Agency Code AAR-500

15. Supplementary Notes
The FAA's Contracting Officer's Technical Representatives (COTR) are Alan Novakoff and Paul Jankowski

16. Abstract

This study was conducted under the auspices of the National Materials Advisory Board's Committee on Commercial Aviation Security to assess the operational performance of explosives detection systems and hardened cargo containers in airports and compare that performance to their performance in laboratory testing, with a focus on ways to deploy these systems more effectively to improve aviation security.

17. Key Words	18. Distribution Statement
explosives detection systems, aviation security, explosives detection technology, security procedures	

19. Security Classif. (of this report)	20. Security Classif. (of this page)	21. No. of Pages	22. Price
Unclassified	Unclassified	68	

Form DOT F1700.7 (8ñ72) Reproduction of completed page authorized